Acclaim for

Randolph M. Nesse and George C. Williams's

WHY WE GET SICK

"This is the most important book written about issues in biomedi-cine in the last fifty years. When the world's leading evolutionary biologist (Williams) teams up with a thoughtful physician (Nesse), the product is a gripping exploration of why our bodies respond the way they do to injury and disease."

—Michael S. Gazzaniga, Ph.D.,
director, Center for Neuroscience,
University of California at Davis

"Darwinian medicine . . . holds that there are evolutionary expla-nations for human disease and physical frailties, just as for everything else in biology, and that these insights can inspire better treatments. . . . In *Why We Get Sick* . . . two proponents of Darwinian medicine lay out the ambitious reach of the adventurous new discipline."

—*The New York Times Magazine*

"Every so often, a book comes along that has the power to change the way we live and die. This splendid book is one, and it could well revolutionize the way physicians are taught, the way they practice, and even the way parents watch over their child with a fever or a cough."

—Professor Robert Ornstein,
author of *The Psychology of Consciousness*

"Would you accept that eating certain kinds of red meat could help ward off heart attacks? That taking aspirin when you are sick could make things worse? That mothers should sleep right next to their infants to prevent sudden infant death? You might after hearing how your prehistoric ancestors lived, according to a small but growing tribe of 'Darwinian medicine' thinkers. They argue that for too long physicians have ignored the forces that shaped us over evolutionary eons. . . . Such ideas are . . . controversial, but that's the point."

—*Wall Street Journal*

"*Why We Get Sick* is certain to be recognized as one of the most important books of the decade, and what's more, it's beautifully written."

—Roger Lewin,
author of *Human Evolution*, 3rd Edition

"*Why We Get Sick* offers both a provocative challenge to medicine and a thoughtful discussion of how evolutionary theory applies to people."

—*Business Week*

Randolph M. Nesse, M.D.
George C. Williams, Ph.D.

WHY WE GET SICK

Randolph M. Nesse, M.D., is a practicing physician and professor and associate chair for education and academic affairs in the Department of Psychiatry at the University of Michigan Medical School.

George C. Williams, Ph.D., is a professor emeritus of ecology and evolution at the State University at Stony Brook and editor of *The Quarterly Review of Biology*.

WHY WE GET SICK

*The New Science
of Darwinian Medicine*

Randolph M. Nesse, M.D.
George C. Williams, Ph.D.

VINTAGE BOOKS

A Division of Random House, Inc.

New York

The Library of Congress has cataloged
the Times Books edition as follows:
Nesse, Randolph M.
Why we get sick : the new science of Darwinian medicine
Randolph M. Nesse and George C. Williams.—1st ed.
p. cm.
Includes bibliographical references and index.
ISBN 0-8129-2224-7
1. Medicine—Philosophy. 2. Human evolution.
3. Human biology. 4. Adaptation (Physiology)
I. Williams, George C. (George Christopher), 1926–   II. Title.
R723.N387 1995
610'.1-dc20 94-27651
Vintage ISBN: 0-679-74674-9

Illustrations by Jared M. Brown

B987

ACKNOWLEDGMENTS

Our work has benefited enormously from comments made by many colleagues and friends who know more than we do about certain aspects of medicine and evolution. We have not always had the sense to take their advice, so don't blame them for our mistakes. Among those who have offered comments or other suggestions on the manuscript are: James Abelson, M.D., Ph.D., Laura Betzig, Ph.D., Helena Cronin, Ph.D., Lyubica Dabich, M.D., Wayne Davis, Ph.D., William Ensminger, M.D., Paul Ewald, Ph.D., Joseph Fantone, M.D., Rosalind Fantone, R.N., Robert Fekety, M.D., Linda Garfield, M.D., Robert Green, M.D., Daniel Hrdy, M.D., Sarah Hrdy, Ph.D., Matt Kluger, Ph.D., Isaac Marks, M.D., Steven Myers, M.D., James Neel, M.D., Ph.D., Margie Profet, M.A., Robert Smuts, M.A., William Soloman, M.D., Paul Turke, Ph.D., Alan Weder, M.D., Brant Wenegrat, M.D., and Elizabeth Young, M.D. For help in finding references we especially thank Doris Williams, Jeanette Underhill, M.D., and Joann Tobin. A sabbatical provided by The University of Michigan with support from John Greden, M.D., and George Curtis, M.D., made it possible for Randolph Nesse to work on the manuscript at Stanford University, where Brant Wenegrat, M.D., and Anne O'Reilly offered hospitality beyond measure. Barbara Polcyn's loyal and effective secretarial support was wonderful. We are grateful to our agent, John Brockman, for convincing us that we could present serious new science in a book for a general audience and for handling negotiations and publishing details with great effectiveness, and to Barbara Williams for persuading us to take John Brockman seriously. The style and structure of the book are much improved thanks to detailed editing by Margaret Nesse and by our editor at Times Books, Elizabeth Rapoport.

Our greatest debt is to those who made us realize that we had a reason to write this book. They are the pioneers and visionaries whose ideas and investigations form the heart of the now flourishing field of Darwinian medicine. Some, like Paul Ewald and Margie Profet, figure prominently in several places in our text. Others are mentioned more briefly or merely have their publications listed in our endnotes. We are confident that, over the next few years, they will all be getting growing shares of the recognition they richly deserve.

CONTENTS

PREFACE

We first met and discovered our shared interests in 1985 at a meeting of a group that later developed into the Human Behavior and Evolution Society. One of us (Nesse) was a physician in the Department of Psychiatry at the University of Michigan Medical School. Frustration with psychiatry's lack of theoretical foundation and fascination with the extraordinary progress that evolutionary ideas had brought to the field of animal behavior had led to his association with the University of Michigan Evolution and Human Behavior Program. Colleagues in that interdisciplinary group, on hearing about his long-term interest in the evolutionary origins of aging, suggested a 1957 paper by a biologist named George Williams. The paper was a revelation. Aging had an evolutionary explanation. Why not anxiety disorders or schizophrenia? Thanks to subsequent years of conversations with evolutionists, especially Williams, and with medical school residents and faculty, he has found that an evolutionary perspective on patients' disorders has become steadily more natural and useful.

The other author (Williams) has divided his career between marine ecological research and theoretical studies on evolution. His interest in medical applications of evolutionary ideas was aroused by reading a 1980 article by Paul Ewald in *The Journal of Theoretical Biology*, "Evolutionary Biology and the Treatment of the Signs and Symptoms of Infectious Disease." Ewald's work suggested that evolutionary ideas might well have significance for many medical problems, not just those that arise from infection. Williams' general knowledge of evolutionary genetics included many principles with obvious implications for genetic diseases, and his early work on the evolution of the aging process suggested a basic relevance of evolution to gerontology.

We convinced each other, shortly after we met, that the potential contribution of evolutionary biology to medical progress was important enough to justify a real effort to bring this idea to others. We decided to put our reasoning and some obvious examples into print as a way of stimulating other workers to explore many other possibilities. After our jointly written article, "The Dawn of Darwinian Medicine," published in *The Quarterly Review of Biology* in March 1991, drew a favorable reception from the press as well as colleagues in both medicine and evolutionary biology, we decided that it could easily be expanded into a book that would interest a wide range of readers.

Charles Darwin's theory of natural selection as the explanation for the functional design of organisms is the foundation of almost everything in this book. The discussion centers on the concept of adaptation by natural selection: adaptations by which we combat pathogens, adaptations of pathogens that counter our adaptations, maladaptive but necessary costs of our adaptations, maladaptative mismatches between our body's design and our current environments, and so on.

As we wrote, we kept discovering new ways in which Darwinism can aid the progress of medicine. We gradually realized that Darwinian medicine is not just a few ideas, but a whole new field, with exciting new developments arising at an ever-increasing rate. However, we must emphasize that Darwinian medicine is still in its infancy. The examples of Darwinian thinking applied to medical problems should not be taken as authoritative conclusions or medical advice. They are designed only to illustrate the use of evolutionary thinking in medicine, *not* to instruct people on how to protect their health or treat their diseases. This is not to say that we believe Darwinian medicine is merely a theoretical endeavor. Far from it! We have every expectation that the pursuit of evolutionary questions will demonstrably improve human health. That will require effort, money, and time. In the meanwhile, we hope this book will stimulate people to think about their illnesses in a different way, to ask questions of their doctors, perhaps even argue with them, but certainly not to ignore their instructions.

Having made that disclaimer, we will also make a few others. This book does not arise from a disapproval of current medical research or practice in Western industrialized nations. It is based on the conviction that medical research and practice would be even better if questions of adaptation and historical causation were routinely considered along with those of immediate physical and chemical causation. We are urging not an alternative to modern medical practice but rather an additional perspective from a well-established body of scientific knowledge that has been largely neglected by the medical profession. We would be very much against Darwinian medicine being viewed as a kind of alternative cult opposed to some supposed orthodoxy. It is likewise not our purpose to make political recommendations, although we believe that some of our reasoning might prove important to those who formulate health care or environmental policies.

In addition to trying to make this book interesting and informative to a wide audience, we have tried to make it a preliminary but scientifically valid guide for physicians and researchers who are asking evolutionary questions in their own areas of expertise. We well realize that many medical professionals have already been asking such questions. Often, however, they have done so apologetically, treating their own ideas not as serious hypotheses but as mere speculations undeserving of serious inquiry. We challenge this attitude as strongly as possible and hope that the examples in this book will make many scientists realize that their evolutionary hypotheses are legitimate and deserve scientific testing, in ways that may be easier and more decisive than they suspect. This book does not offer formal instruction on how to test evolutionary hypotheses, but it does give many examples of such testing.

We hope readers will realize that this meager book can provide only a brief glimpse of a few current evolutionary ideas in relation to a select list of medical examples. Medicine is now such a huge field that no one can master more than a small part of it. Even specialties such as internal medicine are quickly splitting into subspecialties, such as cardiology, and into subsubspecialties. Neither of us claims to have mastered more than a small fraction of the knowledge encompassed by modern medicine. We are well aware that any discussion of such a wide range of topics as is found in this book must of necessity be superficial and oversimplified. We hope that this will not seriously mislead anyone and that specialists will forgive us for any minor inaccuracies they may find. These risks seem worth it because of the potential utility of a broad overview of Darwinian medicine and because we believe that readers will derive real pleasure from an evolutionary understanding of their bodies' functioning, and occasional malfunctioning.

WHY WE GET SICK

1

THE
MYSTERY OF
DISEASE

Why, in a body of such exquisite design, are there a thousand flaws and frailties that make us vulnerable to disease? If evolution by natural selection can shape sophisticated mechanisms such as the eye, heart, and brain, why hasn't it shaped ways to prevent nearsightedness, heart attacks, and Alzheimer's disease? If our immune system can recognize and attack a million foreign proteins, why do we still get pneumonia? If a coil of DNA can reliably encode plans for an adult organism with ten trillion specialized cells, each in its proper place, why can't we grow a replacement for a damaged finger? If we can live a hundred years, why not two hundred?

We know more and more about why individuals get specific diseases but still understand little about why diseases exist at all. We know that a high-fat diet causes heart disease and sun exposure causes skin cancer, but why do we crave fat and sunshine despite their dangers? Why can't our bodies repair clogged arteries and sun-damaged skin? Why does sunburn hurt? Why does anything hurt? And why are we, after millions of years, still prone to streptococcal infection?

The great mystery of medicine is the presence, in a machine of exquisite design, of what seem to be flaws, frailties, and makeshift mechanisms that give rise to most disease. An evolutionary approach transforms this mystery into a series of answerable questions: Why hasn't the Darwinian process of natural selection steadily eliminated the genes that make us susceptible to disease? Why hasn't it selected for genes that would perfect our ability to resist damage and enhance repairs so as to eliminate aging? The common answer—that natural selection just isn't powerful enough—is usually wrong. Instead, as we will see, the body is a bundle of careful compromises.

The body's simplest structures reveal exquisite designs unmatched by any human creations. Take bones. Their tubular form maximizes strength and flexibility while minimizing weight. Pound for pound, they are stronger than solid steel bars. Specific bones are masterfully shaped to serve their functions—thick at the vulnerable ends, studded with surface protrusions where they increase muscle leverage, and grooved to provide safe pathways for delicate nerves and arteries. The thickness of individual bones increases wherever strength is needed. Wherever they bend, more bone is deposited. Even the hollow space inside the bones is useful: it provides a safe nursery for new blood cells.

Physiology is still more impressive. Consider the artificial kidney machine, bulky as a refrigerator yet still a poor substitute that performs only a few of the functions of its natural counterpart. Or take the best man-made heart valves. They last only a few years and crush some red blood cells with each closure, while natural valves gently open and close two and a half billion times over a lifetime. Or consider our brains, with their capacity to encode the smallest details of life that, decades later, can be recalled in a fraction of a second. No computer can come close.

The body's regulatory systems are equally admirable. Take, for instance, the scores of hormones that coordinate every aspect of life, from appetite to childbirth. Controlled by level upon level of feedback loops, they are far more complex than any man-made chemical factory. Or consider the intricate wiring of the sensorimotor system. An image falls onto the retina; each cell transmits its signal via the optic nerve to a brain center that decodes shape, color, and movement, then to other brain centers that link with memory banks to determine that the image is that of a snake, then to fear centers and decision centers that motivate and initiate action, then to motor nerves that contract exactly the right muscles to jerk the hand away—all this in a fraction of a second.

Bones, physiology, the nervous system—the body has thousands of consummate designs that elicit our wonder and admiration. By contrast, however, many aspects of the body seem amazingly crude. For instance, the tube that carries food to the stomach crosses the tube that carries air to the lungs, so that every time we swallow, the airway must be closed off lest we choke. Or consider nearsightedness. If you are one of the unlucky 25 percent who have the genes for it, you are almost certain to become nearsighted and thus unlikely to recognize a tiger until you are nearly its dinner. Why haven't these genes been eliminated? Or take atherosclerosis. An intricate network of arteries carries just the right amount of blood to every part of the body. Yet many of us develop cholesterol deposits on the walls of our arteries, and the resulting blockage in blood flow causes heart attacks and strokes. It is as if a Mercedes-Benz designer specified a plastic soda straw for the fuel line!

Dozens of other bodily designs seem equally inept. Each may be considered a medical mystery. Why do so many of us have allergies? The immune system is useful, of course, but why can't it leave pollen alone? For that matter, why does the immune system sometimes attack our own tissues to cause multiple sclerosis, rheumatic fever, arthritis, diabetes, and lupus erythematosus? And then there is nausea in pregnancy. How incomprehensible that nausea and vomiting should so often plague future mothers at the very time when they are assuming the burden of nourishing their developing babies! And how are we to understand aging, the ultimate example of a universal occurrence that seems functionally incomprehensible?

Even our behavior and emotions seem to have been shaped by a prankster. Why do we crave the very foods that are bad for us but have less desire for pure grains and vegetables? Why do we keep eating when we know we are too fat? And why is our willpower so weak in its attempts to restrain our desires? Why are male and female sexual responses so uncoordinated, instead of being shaped for maximum mutual satisfaction? Why are so many of us constantly anxious, spending our lives, as Mark Twain said, "suffering from tragedies that never occur"? Finally, why do we find happiness so elusive, with the achievement of each long-pursued goal yielding not contentment, but only a new desire for something still less attainable? The design of our bodies is simultaneously extraordinarily precise and unbelievably slipshod. It is as if the best engineers in the universe took every seventh day off and turned the work over to bumbling amateurs.

TWO KINDS OF CAUSES

To resolve this paradox, we must discover the evolutionary causes for each disease. By now it is obvious that these evolutionary causes of disease are different from the causes most people think of. Consider heart attacks. Eating fatty foods and having genes that predispose to atherosclerosis are major causes of heart attacks. These are what biologists call *proximate* ("near") causes. We are more interested here in the *evolutionary* causes, those that reach further back to why we are designed the way we are. In studying heart attacks, the evolutionist wants to know why natural selection hasn't eliminated the genes that promote fat craving and cholesterol deposition. Proximate explanations address how the body works and why some people get a disease and others don't. Evolutionary explanations show why humans, in general, are susceptible to some diseases and not to others. We want to know why some parts of the human body are so prone to failure, why we get some diseases and not others.

When proximate and evolutionary explanations are carefully distinguished, many questions in biology make more sense. A proximate explanation describes a trait—its anatomy, physiology, and biochemistry, as well as its development from the genetic instructions provided by a bit of DNA in the fertilized egg to the adult individual. An evolutionary explanation is about why the DNA specifies the trait in the first place and why we have DNA that encodes for one kind of structure and not some other. Proximate and evolutionary explanations are not alternatives—both are needed to understand every trait. A proximate explanation for the external ear would include information about how it focuses sound, the tissues it is made of, its arteries and nerves, and how it develops from the embryo to the adult form. Even if we know all this, however, we still need an evolutionary explanation of how its structure gives creatures with ears an advantage, why those that lack the structure are at a disadvantage, and what ancestral structures were gradually shaped by natural selection to give the ear its current form. To take another example, a proximate explanation of taste buds describes their structure and chemistry, how they detect salt, sweet, sour, and bitter, and how they transform this information into impulses that travel via neurons to the brain. An evolutionary explanation of taste buds shows why they detect saltiness, acidity, sweetness, and bitterness instead of other chemical characteristics, and how the capacities to detect these characteristics help the bearer to cope with life.

Proximate explanations answer "what?" and "how?" questions about structure and mechanism; evolutionary explanations answer "why?" questions about origins and functions. Most medical research seeks proximate explanations about how some part of the body works or how a disease disrupts this function. The other half of biology, the half that tries to explain what things are for and how they got there, has been neglected in medicine. Not entirely, of course. A primary task of physiology is to find out what each organ normally does; the whole field of biochemistry is devoted to understanding how metabolic mechanisms work and what they are for. But in clinical medicine, the search for evolutionary explanations of disease has been halfhearted at best. Since disease is often assumed to be necessarily abnormal, the study of its evolution may seem preposterous. But an evolutionary approach to disease studies not the evolution of the disease but the design characteristics that make us susceptible to the disease. The apparent flaws in the body's design, like everything else in nature, can be fully understood only with evolutionary as well as proximate explanations.

Are evolutionary explanations mere speculations, of intellectual interest only? Not at all. For instance, consider morning sickness. If, as Seattle researcher Margie Profet has suggested, the nausea, vomiting, and food aversions that often accompany early pregnancy evolved to protect the developing fetus from toxins, then the symptoms should begin when fetal-tissue differentiation begins, should decrease as the fetus becomes less vulnerable, and should lead to avoidance of foods that contain the substances most likely to interfere with fetal development. As we will see, substantial evidence matches these predictions.

Evolutionary hypotheses thus predict what to expect in proximate mechanisms. For instance, if we hypothesize that the low iron levels associated with infection are not a cause of the infection but a part of the body's defenses, we can predict that giving a patient iron may worsen the infection—as indeed it can. Trying to determine the evolutionary origins of disease is much more than a fascinating intellectual pursuit; it is also a vital yet underused tool in our quest to understand, prevent, and treat disease.

THE CAUSES OF DISEASE

Experts on various diseases often ask themselves why a particular disease exists at all, and they often have some good ideas. In many cases, however, they confuse evolutionary with proximate explanations, or do not know how to go about testing their ideas, or are simply reluctant to propose explanations that seem outside the mainstream. These difficulties can perhaps be reduced with the help of a formal framework for Darwinian medicine. To this end, we propose six categories of evolutionary explanations of disease. Each of these will be described at length in later chapters, but this brief overview illustrates the logic of the enterprise and provides an overview of the terrain ahead.

1. Defenses

Defenses are not actually explanations of disease, but because they are so often confused with other manifestations of disease we list them here. A fair-skinned person with severe pneumonia may take on a dusky hue and have a deep cough. These two signs of pneumonia represent entirely different categories, one a manifestation of a defect, the other a defense. The skin is blue because hemoglobin is darker in color when it lacks oxygen. This manifestation of pneumonia is like a clank in a car's transmission. It isn't a preprogrammed response to the problem, it is just a happenstance result with no particular utility. A cough, on the other hand, is a defense. It results from a complex mechanism designed specifically to expel foreign material in the respiratory tract. When we cough, a coordinated pattern of movements involving the diaphragm, chest muscles, and voice box propels mucus and foreign matter up the trachea and into the back of the throat, where it can be expelled or swallowed to the stomach, where acid destroys most bacteria. Cough is not a happenstance response to a bodily defect; it is a coordinated defense shaped by natural selection and activated when specialized sensors detect cues that indicate the presence of a specific threat. It is, like the light on a car's dashboard that turns on automatically when the gas tank is nearly empty, not a problem itself but a protective response to a problem.

This distinction between defenses and defects is not merely of academic interest. For someone who is sick it can be crucial. Correcting a defect is almost always a good thing. If you can do something to make the clanking in the transmission stop or the pneumonia patient's skin turn warm pink, it is almost always beneficial. But eliminating a defense by blocking it can be catastrophic. Cut the wire to the light that indicates a low fuel supply, and you are more likely to run out of gas. Block your cough excessively, and you may die of pneumonia.

2. Infection

Given that some bacteria and viruses treat us mainly as meals, we can think of them as enemies. Unfortunately, they are not just simple pests put here to bedevil us but sophisticated opponents. We have evolved defenses to counter their threats. They have evolved ways to overcome our defenses or even to use them to their own benefit. This endlessly escalating arms race explains why we cannot eradicate all infections and also explains some autoimmune diseases. We expand greatly on these topics in the next two chapters.

3. Novel Environments

Our bodies were designed over the course of millions of years for lives spent in small groups hunting and gathering on the plains of Africa. Natural selection has not had time to revise our bodies for coping with fatty diets, automobiles, drugs, artificial lights, and central heating. From this mismatch between our design and our environment arises much, perhaps most, preventable modern disease. The current epidemics of heart disease and breast cancer are tragic examples.

4. Genes

Some of our genes are perpetuated despite the fact that they cause disease. Some of their effects are "quirks" that were harmless when we lived in a more natural environment. For instance, most of the genes that predispose to heart disease were harmless until we began overindulging on fatty diets. The genes that cause nearsightedness cause problems only in cultures where children do close work early in life. Some of the genes that cause aging were subject to little selection when average life spans were shorter.

Many other genes that cause disease have actually been selected for because they provide benefits, either to the bearer or to other individuals with the gene in other combinations. For instance, the gene that causes sickle-cell disease also prevents malaria. In addition to this well-known example, many others are discussed in later chapters, including sexually antagonistic genes that benefit fathers at the expense of mothers or vice versa.

Our genetic code is constantly being disrupted by mutations. On very rare occasions these changes in DNA are beneficial, but much more commonly they create disease. Such damaged genes are constantly being eliminated or kept to a minimum by natural selection. For this reason defective genes with no compensating benefit are not a common cause of disease.

Finally, there are "outlaw" genes that facilitate their own transmission at the expense of the individual and thus bluntly demonstrate that selection acts ultimately to benefit genes, not individuals or species. Because selection among individuals is a potent evolutionary force, outlaw genes are also an uncommon cause of disease.

5. Design Compromises

Just as there are costs associated with many genes that offer an overall benefit, there are costs associated with every major structural change preserved by natural selection. Walking upright gives us the ability to carry food and babies, but it predisposes us to back problems. Many of the body's apparent design flaws aren't mistakes, just compromises. To better understand disease, we need to understand the hidden benefits of apparent mistakes in design.

6. Evolutionary Legacies

Evolution is an incremental process. It can't make huge jumps, only small changes, each of which must be immediately beneficial. Major changes are difficult to accomplish even for human engineers. Fires occurred when a popular line of pickup truck was struck from the side because the gasoline tanks were located outside the frame. But to locate the tanks within the frame would require a major redesign of everything now there, which could cause new problems and require new compromises. Even human engineers can be constrained by historical legacies. Similarly, our food passes through a tube in front of the windpipe, and must cross it to get to the stomach, thus exposing us to the danger of choking. It would be more sensible to relocate the nostrils to somewhere on the neck, but that will never happen, as we explain in Chapter 9.

WHAT WE ARE NOT SAYING

Before we discuss the details of the above causes of disease, we would like to try to forestall several potentially dangerous misunderstandings. First of all, our enterprise has nothing to do with eugenics or Social Darwinism. We are not interested here in whether the human gene pool is getting better or worse, and we are emphatically not advocating actions to improve the species. We are not even particularly interested in most genetic differences between people, but much more in the genetic material that we all have in common.

An evolutionary perspective on disease does not change the ancient goals of medicine carved on a statue honoring physician E. L. Trudeau's work at Saranac Lake: "To cure, sometimes, To help, often, To console, always." The goal of medicine has always been (and, in our belief, always should be) to help the sick, not the species. Confusion regarding this point has justified much mischief. At the beginning of the century, Social Darwinist ideology helped to justify withholding medical care from the poor and letting capitalist giants battle irrespective of effects on individuals. These beliefs were intimately linked to those of the eugenicists, who advocated sterilization of certain groups in order to improve the species (or race!). Such ideology has long ago earned a well-deserved ill repute. It made metaphorical use of some of the terminology of Darwinism but no use of the theory as biologists understand it. We are by no means advocating that medicine should assist natural selection, nor do we suggest that biology can guide moral decisions. We would never argue that any disease is good, even though we will offer many examples in which pathology is associated with some unappreciated benefit. Darwinism gives no moral guidelines about how we should live or how doctors should practice medicine. A Darwinian perspective on medicine can, however, help us to understand the evolutionary origins of disease, and this knowledge will prove profoundly useful in achieving the legitimate goals of medicine.

2

EVOLUTION BY NATURAL SELECTION

> Now, as each of the parts of the body, like every
> other instrument, is for the sake of some purpose,
> viz. some action, it is evident that the body as a
> whole must exist for the sake of some complex
> action.
>
> —Aristotle

The solutions to the mysteries discussed in Chapter 1 are to be found in the workings of natural selection. The process is fundamentally very simple: natural selection occurs whenever genetically influenced variation among individuals affects their survival and reproduction. If a gene codes for characteristics that result in fewer viable offspring in future generations, that gene is gradually eliminated. For instance, genetic mutations that increase vulnerability to infection, or cause foolish risk taking or lack of interest in sex, will never become common. On the other hand, genes that cause resistance to infection, appropriate risk taking, and success in choosing fertile mates are likely to spread in the gene pool, even if they have substantial costs.

A classic example is the spread of a gene for dark wing color in a British moth population living downwind from major sources of air pollution. Pale moths were conspicuous on smoke-darkened trees and easily caught by birds, while a rare mutant form of moth whose color more closely matched that of the bark escaped the predators' beaks. As the tree trunks became darker, the mutant gene spread rapidly and largely displaced the gene for pale wing color. That is all there is to it. Natural selection involves no plan, no goal, and no direction—just genes increasing and decreasing in frequency depending on whether individuals with those genes have, relative to other individuals, greater or lesser reproductive success.

The simplicity of natural selection has been obscured by many misconceptions. For instance, Herbert Spencer's nineteenth-century catch phrase "survival of the fittest" is widely thought to summarize the process, but it actually promotes several misunderstandings. First of all, survival is of no consequence in and of itself. This is why natural selection has created some organisms, such as salmon and annual plants, that reproduce only once, then die. Survival increases fitness only insofar as it increases later reproduction. Genes that increase lifetime reproduction will be selected for even if they result in reduced longevity. Conversely, a gene that decreases total lifetime reproduction will obviously be eliminated by selection even if it increases an individual's survival.

Further confusion arises from the ambiguous meaning of "fittest." The fittest individual, in the biological sense, is not necessarily the healthiest, strongest, or fastest. In today's world, and many of those of the past, individuals of outstanding athletic accomplishment need not be the ones who produce the most grandchildren, a measure that should be roughly correlated with fitness. To someone who understands natural selection, it is no surprise that parents are so concerned about their children's reproduction.

A gene or an individual cannot be called "fit" in isolation but only with reference to a particular species in a particular environment. Even in a single environment, every gene involves compromises. Consider a gene that makes rabbits more fearful and thereby helps to keep them from the jaws of foxes. Imagine that half of the rabbits in a field have this gene. Because they do more hiding and less eating, these timid rabbits might be, on average, a bit less well fed than their bolder companions. If, hunkered down in the March snow waiting for spring, two thirds of them starve to death while this is the fate of only one third of the rabbits who lack the gene for fearfulness, then, come spring, only a third of the rabbits will have the gene for fearfulness. It has been selected against. It might be nearly eliminated by a few harsh winters. Milder winters or an increased number of foxes could have the opposite effect. It all depends on the *current* environment.

NATURAL SELECTION BENEFITS GENES, NOT GROUPS

Many people have seen the nature film in which droves of starving lemmings jump eagerly to a watery death as a resonant voice explains that when food becomes scarce, some lemmings sacrifice themselves so that there will be enough food for at least some of the group to survive. A few decades ago, such "group selection" explanations were taken seriously by professional biologists, but not now. To see why, compare two imaginary lemmings. One is a noble fellow who, upon sensing that the population is about to outrun its food supply, quickly jumps to his death in the nearest stream. The other is a selfish lout who waits for the noble ones to do away with themselves and then eats as much food as he can get, mates as often as possible, and has as many offspring as possible. What would happen to the genes that code for the behavior of sacrificing oneself for the benefit of the group? No matter how beneficial they might be for the species, they would be eliminated.

So how can we explain the observations of apparently suicidal lemmings? When food becomes scarce in late winter, lemmings migrate, rushing along in large groups that do not always stop when they encounter waters created by early snowmelt. Drownings are, however, rather uncommon. To get the footage they wanted, the makers of the film apparently had to use brooms to surreptitiously herd the lemmings into the water, a dramatic example of the human preference for altering reality rather than theory when the two conflict! There are special circumstances in which selection at the group level can outweigh the usually stronger force of selection at the level of the individual, but they do not apply very often.

As British biologist Richard Dawkins, author of *The Selfish Gene*, has emphasized, individuals may be viewed as vessels created by genes for the replication of genes, to be discarded when the genes are through with them. This perspective mightily shakes the common view that evolution tends toward a world of health, harmony, and stability. It does not create such a world. We would like to imagine that life is naturally happy and healthy, but natural selection cares not a whit for our happiness, and it promotes health only when it is in the interests of our genes. If tendencies to anxiety, heart failure, nearsightedness, gout, and cancer are somehow associated with increased reproductive success, they will be selected for and we will suffer even as we "succeed," in the purely evolutionary sense.

KIN SELECTION

We have implied that reproduction is the essence of the fitness maximized by natural selection, and in our discussion of lemmings we indicated that evolution does not favor individuals who act to help others at their own expense. These generalizations tell only part of the story. Ultimately, it is the genetic representation in future generations that counts, whether that is accomplished by having children or by doing things that increase the reproduction of your close relatives, many of whose genes are identical to yours.

Half of the genes in a child are identical to those in the mother, and half are identical to those in the father. Full siblings, on average, also share half of each other's genes. One fourth of the genes in a grandparent are identical to those in the grandchild. Cousins share one eighth of their genes. This means that, from the perspective of your genes, your sister's survival and reproduction are half as important as your own and your cousin's one eighth as important. For this reason, selection favors extending help to relatives if, all else being equal (e.g., age and health), the cost to oneself of extending the help is less than the benefit to the relative times the degree of relationship. In a classic anecdote, British biologist J. B. S. Haldane was asked if he would sacrifice his life for his brother. "No," he said, "not for one brother. But I would for two brothers. Or eight cousins." Formal recognition of this principle and its importance in explaining cooperation awaited the landmark 1964 paper by British biologist William Hamilton, winner of the 1993 Craoord Prize, created to honor scientists whose work is in fields not covered by the Nobel Prize. Another great British biologist, John Maynard Smith, christened the phenomenon *kin selection*.

Another apparent exception to the nice-guys-finish-last principle in evolution is the result of reciprocal exchanges of favors between individuals who need not be relatives. If Elsa is an expert maker of shoes and Fritz is a skillful hunter of animals that supply excellent leather, trading resources will benefit them both. It pays me to be nice to you, and vice versa. Ever since Robert Trivers's classic 1971 paper on reciprocity theory, biologists have routinely interpreted cooperative relations among organisms in nature as resulting from either reciprocal exchanges or kin selection. The biology of social life has grown thanks to the efforts of pioneers such as E. O. Wilson, author of *Sociobiology*, and Richard Alexander, author of *Darwinism and Human Affairs*. Early controversies and misunderstandings have been largely supplanted by growing work in this new field of science.

HOW DOES NATURAL SELECTION OPERATE?

There is a widespread misconception that evolution proceeds according to some plan or direction, but it has neither, and the role of chance ensures that its future course will be unpredictable. Random variations in individual organisms create tiny differences in their Darwinian fitness. Some individuals have more offspring than others, and the characteristics that increased their fitness thereby become more prevalent in future generations. Once upon a time (at least) a mutation occurred in a human population in tropical Africa that changed the hemoglobin molecule in a way that provided resistance to malaria. This enormous advantage caused the new gene to spread, with the unfortunate consequence that sickle-cell anemia came to exist, as will be discussed in later chapters.

Chance can influence the outcome at each stage: first in the creation of a genetic mutation; second in whether the bearer lives long enough to show its effects; third in chance events that influence the individual's actual reproductive success; fourth in whether a gene, even if favored in one generation, is, by happenstance, eliminated in the next; and finally in the many unpredictable environmental changes that will undoubtedly occur in the history of any group of organisms. As Harvard biologist Stephen Jay Gould has so vividly expressed it, if one could rewind the tape of biological history and start the process over again, the outcome would surely be different. Not only might there not be humans, there might not even be anything like mammals.

We will often emphasize the elegance of traits shaped by natural selection, but the common idea that nature creates perfection needs to be analyzed carefully. The extent to which evolution achieves perfection depends on exactly what you mean. If you mean "Does natural selection always take the best path for the long-term welfare of a species?" the answer is no. That would require adaptation by group selection, and this is, as noted above, unlikely. If you mean "Does natural selection create every adaptation that would be valuable?" the answer again is no. For instance, some kinds of South American monkeys can grasp branches with their tails. This trick would surely also be useful to some African species, but, simply because of bad luck, none have it. Some combination of circumstances started some ancestral South American monkeys using their tails in ways that ultimately led to an ability to grab onto branches, while no such development took place in Africa. Mere usefulness of a trait does not necessarily mean that it will evolve.

There is a sense, however, in which natural selection does regularly come close to perfection, and that is in optimizing some quantitative features. If a trait serves a specific function, selection among minor modifications over many generations tends to make its quantitative aspects closely approach the functional ideal. For instance, a bird's wings must be long enough to give good lift but short enough to allow the bird to maintain control. Measurements on birds found killed after a major storm showed more than expected numbers of unusually long or unusually short wings. The survivors showed a bias toward intermediate (more nearly optimal) wing lengths.

In human physiology, there are hundreds of similar examples in which traits have been shaped to nearly optimal values: the sizes and shapes of bones, blood pressure, glucose level, pulse rate, age at onset of puberty, stomach acidity—the list could go on and on. The observed values may never be exactly perfect, but they usually come close. When we think that natural selection has erred, it is more likely that we have missed some important consideration. For instance, stomach acid aggravates ulcers, yet people who take antacids can still digest their food. So is there too much acid? Probably not, given the importance of stomach acid in digestion and in killing bacteria, including those that cause tuberculosis. To identify the imperfections of the body, one must first understand its perfections and the compromises on which many of them are based.

Like any engineer, evolution must constantly compromise. An auto designer could increase the thickness of the fuel tank in order to decrease the risk of fire, but at some point increased cost and decreased mileage and acceleration require a compromise. Thus, fuel tanks do rupture in some collisions, and this compromise costs some lives each year. While natural selection cannot achieve perfection in every character simultaneously, its compromises are not random but are accurately shaped to give the greatest net benefit.

An apocryphal story tells of Henry Ford looking at a junkyard filled with Model Ts. "Is there anything that never goes wrong with any of these cars?" he asked. Yes, he was told, the steering column never fails. "Well then," he said, turning to his chief engineer, "redesign it. If it never breaks, we must be spending too much on it." Natural selection similarly avoids overdesign. If something works well enough that its deficiencies do not constitute a selective force, there is no way natural selection can improve it. Thus, while every part of the body has some reserve capacity to deal with occasionally encountered extreme circumstances, every part is also vulnerable when its reserve capacity is exceeded. There is nothing in the body that never goes wrong.

Moderate increments of a resource often have enormous value, while higher amounts may have less benefit. If you are making a stew, two onions may be better than one, but ten onions would be much more expensive yet offer little, if any, extra benefit. Such cost-benefit analyses are routine procedures in economics, but they are useful in biology and medicine as well. Consider the use of an antibiotic for pneumonia. A tiny dose will probably have no detectable benefit, a moderate dose will cost more but offer much greater benefits, while a high dose will have still higher costs with no additional benefits and perhaps significant danger.

Just as there are costs as well as benefits involved in every engineering or medical decision, there are costs associated with every beneficial genetic change preserved in evolution. Natural selection isn't weak or capricious; it just selects for genes that give an overall fitness advantage, even if those same genes increase vulnerability to some disease. Is there any way, for instance, for anxiety to be a functionally desirable trait? Consider what would happen to those rabbits we discussed if they had no anxiety in a year when foxes were especially abundant. Even some genes that cause aging are not necessarily maladaptive. They may give benefits during the early years of life, when selection is the strongest, benefits that are more important to fitness than the later costs of aging and inevitable death. To understand disease better, we need to understand the hidden benefits of apparent mistakes in design.

TESTING EVOLUTIONARY HYPOTHESES

This chapter started with a quotation from Aristotle for a serious reason. We can think of him as the originator of the general procedure for functional analysis that has been particularly fruitful in many kinds of biological research and that we expect to be similarly rewarding in medicine. There is, of course, a big difference between Aristotle's outlook and that of modern biologists. He had almost no grasp of the physical and chemical principles that underlie the workings of any organism. He didn't think experiments were necessary. He had no notion of the principle of natural selection and certainly did not realize that organisms were designed entirely to maximize their success in reproduction. Whether applied to the human hand or brain or immune system, Aristotle's powerful question, "What is it for?" now has a very specific scientific meaning: "How has this trait contributed to reproductive success?" His conviction that the body as a whole exists for the sake of some complex action is correct. Only in the past few decades has it become clear that that complex action is reproduction.

Many people have the notion that questions about the function of a trait are not scientific, that they are "teleological" or "speculative" and therefore not appropriate objects of scientific inquiry. This idea is incorrect, as many examples in this book will demonstrate. Questions about the adaptive function of a biological trait are just as amenable to scientific inquiry as are questions about anatomy and physiology. It makes sense to ask about the adaptive significance of biological traits such as eyes, ears, and the cough reflex because they are products of historical processes that have gradually modified them in ways that improve their capacity to serve special functions.

Yet when we ask these "why" questions, we must guard against too readily believing fanciful stories. Why do we have prominent noses? It must be to hold up eyeglasses. Why do babies cry for no apparent reason? It must be to exercise their lungs. Why do we nearly all die by age 100? It must be to make room for new individuals. Almost anything can be the subject of such speculation, but if this is as far as it goes it is not science. The problem is not in the questions but in a lack of adequate investigation and critical thinking about suggested answers.

The above absurd examples demonstrate how easily some explanations can be tested and proven false. Noses could not have evolved to hold up glasses, since we had noses long before we had eyeglasses. Crying cannot be to develop the lungs, since lung health in adulthood does not require crying in infancy. Aging cannot have evolved to make room for new individuals, because natural selection cannot favor such benefits to the group and the details of aging simply do not conform to the expectations for such a function.

Other functional hypotheses are so easily supported that they are of little interest. Anyone thoroughly familiar with the heart's structure and operation can see that it pumps blood. One can also see that coughing expels foreign material from the respiratory tract and that shivering increases body heat. You don't need to be an evolutionary scientist to figure out that teeth allow us to chew food. The interesting hypotheses are those that are plausible and important but not so obviously right or wrong. Such functional hypotheses can lead to new discoveries, including many of medical importance.

THE ADAPTATIONIST PROGRAM

Studies of the functional reasons for human attributes are based on a method of investigation recently named the *adaptationist program*. By suggesting the functional significance of some known aspect of human biology, you may logically be able to predict some other, unknown aspects. An appropriate investigation can then confirm that these characteristics are either there or not. If they are there, they may be of medical significance. If they're not, we can eliminate our hypothesis and go back to the drawing board.

We will give three examples here of interesting discoveries made by considering questions on how various features might contribute to fitness. They relate to beavers and birds but not to medical questions, for which we will give many examples in the chapters to come. To various degrees these examples show that intuitive ideas about fitness, even the intuitions of professional biologists, may not always be adequate. Serious, often mathematical, theorizing is needed to provide the logical answers that can then be tested by investigating real organisms.

Beavers harvest trees in or near their ponds for their food and shelter. They use their teeth to chop through the trunks near the ground, drag the trees to the water if they are not already in it, and tow them to their lodges. How do beavers decide which trees to chop down? They do so *adaptively*, was the hypothesis considered by Michigan biologist Gary Belovsky. This implies an economically rational decision based on a tree's likely value to a beaver, the difficulty expected in chopping it down and moving it, and how far it is from home. Belovsky's calculations showed that an efficient beaver ought to be increasingly discriminating as the distance from the pond increases. Small trees may be rejected for not being worth the time to transport them, large ones for not being worth the labor of felling and transporting them, especially dragging them or pieces of them through the woods to where they can be floated in the pond. Belovsky predicted that the range of sizes of trees harvested by beavers would steadily decrease as the distance from the pond increased. At some point, only trees of an ideal size would be harvested; beyond that, none at all. Observation of stumps of beaver-felled trees near their ponds confirmed the prediction. The next time you see a beaver pond, admire not only the beaver's legendary industry but also its cleverness at setting priorities.

Now imagine a woodland songbird about to lay a clutch of eggs that she and her mate will incubate. Her reproductive success for this breeding season will depend entirely on those eggs. How many should she lay? Remember, she is not trying to assure the survival of the species, she is trying to maximize her own lifetime reproductive success. Laying too few eggs would obviously be foolish, but laying too many can also decrease her total lifetime reproduction if there is not enough food and some of the chicks die, or if she exhausts her energy reserves in caring for her brood and thus jeopardizes her chances of living until the next breeding season. These considerations apply equally to every individual in the woodland, but different birds reach different decisions on how many eggs to lay. If the average for a species is four eggs per pair, some pairs may have five and some only three. Do we conclude that all are trying for four but some can't count? Or do we perhaps conclude that egg numbers are not subject to optimization by natural selection?

An adaptationist forgoes such explanations until after considering the possibility that the birds deserve more credit. Could it be that, as a general rule, three eggs is best for those that lay only three, four for those that lay four, and so on? A simple sort of experiment provides the answer. If there are thirty nests with four eggs, leave ten randomly selected nests alone. From ten other nests remove an egg (the owners are now down to three) and add them to the ten remaining nests (four-egg birds now have five eggs). Now measure the average success of the three groups of birds: those allowed to choose their own egg number and those with one more or one less than they originally laid.

If all relevant factors are carefully considered, the results of such studies usually vindicate the conclusion reached fifty years ago by Oxford ornithologist David Lack: birds adjust the number of eggs they lay to maximize their individual reproductive success. To do this requires an accurate assessment of their own individual health and capabilities and experience. Having to provide food for four nestlings is more difficult and hazardous than providing for only three. Nestlings in more crowded nests may weigh less at fledging and be less likely to survive the following winter. Conditions vary unpredictably from year to year, and worse-than-normal years are especially dangerous for the more crowded broods. Surely such knowledge enhances a naturalist's pleasure in watching a pair of wild birds feed their young. Those birds are doing it right—not just right in general or on average, but right for them as unique individuals.

In this discussion of clutch size we considered the optimal number of offspring. We ignored the fact that there are two kinds of offspring, male and female. Should our birds ideally produce one or the other or both in some ideal proportion? In the natural selection of sex ratio one overwhelmingly important strategy maximizes fitness: producing offspring of whichever sex is in short supply. Any frequenter of singles bars knows that the minority sex has a mating advantage. In nature, individuals that produce male offspring when females are scarce will be selected against because many of those males will never have offspring. If males are scarce, individuals that produce females will not have as many grand-offspring as individuals who produce males. The operation of this process of selection explains why there are equal numbers of males and females. This simple, elegant evolutionary explanation was first recognized by the great evolutionary geneticist

R. A. Fisher in 1930. If you are thinking that an equal sex ratio arises because an individual has an equal chance of getting an X or a Y chromosome from its father, you are right, but this is a proximate explanation. The insufficiency of a proximate explanation is demonstrated by the many special cases such as ants and fig wasps, which are too complex to describe here but in which grossly unequal sex ratios turn out to match the more complex predictions.

Does natural selection in fact produce populations with exactly the same number of males and females? No, it does not, as would be expected by detailed reflection on factors such as the two sexes reaching maturity at different ages, differing death rates, differing costs to male and female parents, and other factors. Careful calculations support the conclusion that, for organisms with sex-determination and reproductive processes like ours, the sex ratio will stabilize when the parents collectively spend equal resources on rearing sons and rearing daughters. The demography of human and many other populations conforms closely to these expectations.

We hope to convince you in the coming chapters that the modern theory of natural selection can be just as helpful in making medically important discoveries as it is for predicting the foraging patterns of beavers, the effects of altered clutch sizes of birds, and the sex ratios of mammals. The reasoning will always start with some prior information about health or disease and a question about evolved adaptation: Is this feature of the human body a part of some adaptive machinery? If so, what must the rest of the machinery be like? How can we test our predictions for unknown aspects of the machinery? If any feature of human biology seems functionally undesirable, how can natural selection have permitted it to arise? Is an undesirable trait the price of a positive feature? Could it be a trait that was adaptive in the Stone Age but that now causes disease? What are the medical consequences of natural selection acting to improve adaptation in our pathogens and parasites? These are just a few of the sorts of questions now routinely asked by evolutionary biologists, and efforts at answering them have been enormously fruitful.

We must temper our enthusiasm with a note of caution. A question about function can have more than one right answer. For instance, the tongue is important both for chewing and for speech; the eyebrows, both for keeping the sweat out of the eyes and for communication. Second, the evolutionary history of a species or a disease is like any other kind of history. There is no experiment, in the usual sense, that we can do now to decide how long ago our ancestors first started to use fires for cooking or other purposes and what subsequent evolutionary effects that change may have had. History can be investigated only by examining the records it has left. Charred bones or even carbon deposits from an ancient campfire can be informative documents to people who know how to read them. Likewise, the chemical structure of proteins and DNA may be read to reveal relationships among now strikingly different organisms. Until a time machine is invented, we will not be able to go back and watch the evolution of major traits, but we can nonetheless reconstruct prehistoric events by the records they left in fossils, carbon traces, structures, and behavioral tendencies, as well as protein and DNA structures. Even when we cannot reconstruct the history of a trait, we can often still be confident that it was shaped by natural selection. This can be supported by evidence for its function in other species and by the match between the trait's characteristics and its functions.

So hypotheses about the evolutionary origins and functions of a trait, just like hypotheses about proximate aspects of a trait, need testing and are often testable. Special difficulties attend the testing of evolutionary hypotheses, but these are no reason to give up—they just make the work more challenging and interesting. Do we claim to test evolutionary hypotheses in this book? Not really. While we will try to separate speculation from fact, and will cite evidence for most of our examples, hardly any of them can be considered proven by the evidence we present. Some of the examples are based on many studies, each with different data bearing on a different aspect of the problem, but even this is often insufficient.

Our goal is not to prove any specific hypothesis but to show that evolutionary questions are interesting, important, and testable. We want people to start asking new questions. So, without apology, we ask questions about the possible evolutionary significance of diverse aspects of disease and offer answers that are often speculative. Some people will, despite our warnings, insist on taking these speculations as facts. Perhaps in a few years Darwinian medicine will have enough confirmed findings to fill a book. For now, our goal is not to exhaustively test a few hypotheses but to encourage patients, doctors, and researchers to ask new questions about why disease exists. As Gertrude Stein said on her deathbed, "The answer, the answer, the answer. What is the answer? . . . In that case, what is the question?"

3

SIGNS AND
SYMPTOMS OF
INFECTIOUS DISEASE

S uppose you are on the side of the mice in cat-mouse conflicts. The mice say they hate the smell of a cat. It makes
them jittery and unable to concentrate on important matters, such as food and courtship and babies. You know of a
drug that will dull the sense of smell so that the mice will no longer be
bothered by the odor of cats. Do you prescribe the drug? Probably
not. The ability to detect cat odor, however unpleasant it may be, is a
valuable asset for mice. The presence of the cat's smell may signal the
imminent arrival of its claws and teeth, and avoiding these is far more
important than the stress of an unpleasant odor.

More realistically, suppose you are a pediatrician treating children
with colds. Colds bring many symptoms that children dislike—
runny nose, headache, fever, and malaise. Acetaminophen (e.g.,
Tylenol) can reduce or eliminate some of these symptoms. Do you
tell the parents of cold-stricken children to give them acetaminophen? If you are a traditional physician or are in the habit of
using acetaminophen yourself to relieve similar symptoms, you
probably do. Is this wise? Consider the analogy between acetaminophen and the drug we were considering for the mice. Like the
smell of a cat, fever is unpleasant but useful. It is an adaptation
shaped by natural selection specifically to fight infection.

FEVER AS DEFENSE AGAINST INFECTION

Matt Kluger, a physiologist at the Lovelace Institute, believes that "there is overwhelming evidence in favor of fever being an adaptive host response to infection that has persisted throughout the animal kingdom for hundreds of millions of years." He believes that using drugs to suppress fever may sometimes make people sicker—and even kill them. Some of the best evidence comes from his laboratory. In one experiment, he showed that even cold-blooded lizards benefit from fever. When infected, they seek out a place warm enough to raise their body temperature about two degrees Celsius. If they cannot move to a warm place, they are more likely to die. Baby rabbits also cannot generate a fever, so when they are sick they too seek out a warm place to raise their body temperature. Adult rabbits do get fever when infected, but if the fever is blocked with a fever-lowering drug, they are more likely to die.

Fever results not from any mistake in temperature regulation but from the activation of a sophisticated evolved mechanism. If you put a rat with a two-degree fever into a very hot room, the rat activates its cooling mechanisms to keep its body temperature two degrees above normal. If you put it into a cooler room, it activates heat-conservation mechanisms to maintain that two-degree fever. Body temperature is carefully regulated even during fever; the thermostat is just set a bit higher.

Perhaps the most dramatic human evidence for the value of fever comes from studies by Julius Wagner-Jauregg in the early decades of this century. After noting that some syphilis patients improved after getting malaria and that syphilis was rare in areas where malaria was common, he intentionally infected thousands of syphilis patients with malaria. In an era when fewer than one in a hundred syphilis patients recovered, this treatment achieved remission rates of 30 percent, an advance that made Wagner-Jauregg worthy of his 1927 Nobel Prize in Physiology or Medicine. At that time, the value of fever was much more widely recognized than it is now.

Doctors still say, as the joke goes, "Take two aspirin and call me in the morning." This isn't so surprising, given that only a few human studies have tried to evaluate fever as an adaptation to combat infection. In one study, children with chicken pox who were given acetaminophen took on average about a day longer to recover than those who took a placebo (sugar pill). In another study, fifty-six volunteers got colds on purpose, from an infectious nasal spray. Some then took aspirin or acetaminophen, others a placebo. The placebo group had a significantly higher antibody response and less nasal stuffiness. They also had a slightly shorter period of infectious dispersal of viruses. The paucity of detailed studies of this sort, given that so many drugs are used to relieve the symptoms of so many infectious diseases in so many patients, shows the reluctance to study the adaptive aspects of unpleasant symptoms.

This may be about to change. Dr. Dennis Stevens, professor of medicine at the University of Washington, cites "evidence that treating a fever in certain circumstances actually may make it more likely the patient will develop septic shock." Medications that block fever apparently interfere with the normal mechanisms that regulate the body's response to infection, with results that may be fatal.

Before going on to other defenses, we should emphasize that a given expression of a defense need not be adaptive, and that even when it is, it may not be essential. We would not dream of recommending that people never take drugs to reduce fever. Even if many studies were to establish decisively that fever is usually important for combating infection, that would not justify an unbending policy of encouraging fever or even of routinely letting it rise to its natural level. An evolutionary perspective draws attention to the costs as well as the benefits of an adaptation like fever. If there were no compensating disadvantage in having the human body operate at 40° C. (103° F.), it ought to stay at that temperature all the time, so as to prevent infections from ever getting started. But even this moderate fever has costs; it depletes nutrient reserves 20 percent faster and causes temporary male sterility. Still higher fevers can cause delirium and perhaps seizures and lasting tissue damage. It should also be realized that no regulation mechanism can perfectly anticipate all situations. We would expect temperature to rise, on average, to a level close to an optimum to fight infection, but because regulatory precision is limited, fever will sometimes rise too much and at other times not enough.

Even if we knew that it would prolong an infection, we would still sometimes want to block fever. Maintaining and improving health are, after all, not the only goals of medicine. If she is about to sing Nanetta in a Metropolitan Opera performance of *Falstaff*, soprano

Barbara Bonney might well decide to take a medication to relieve a touch of laryngitis, even if she knew it might delay her complete recovery. The rest of us may choose to take drugs just to feel better during a cold, even though our recovery might be slower.

The important point, with respect to the adaptive significance of fever, is that we need to know what we are doing before we interfere with it. At present we don't. If discomfort were the whole story, we could always choose to reduce or eliminate it. But if reducing fever will often delay recovery or increase the likelihood of secondary infection, we should interfere only when the expected gain is worth the risk. We hope that medical research will soon produce the evidence to help doctors and patients decide when fever is and is not useful.

IRON WITHHOLDING

Our bodies have a related defense mechanism, of which most people are unaware and which physicians sometimes unwittingly attempt to frustrate. Here are some clues about how it works. A patient with chronic tuberculosis is found to have a low level of iron in his blood. A physician concludes that correcting the anemia may increase the patient's resistance, so she gives him an iron supplement. The patient's infection gets worse. Another clue: Zulu men often drink beer made in iron pots and often get serious liver infections caused by an amoeba. In contrast, less than 10 percent of Masai tribesmen have amoebic infections. They are herdsmen and drink large amounts of milk. When a group of Masai were given iron supplements, 88 percent soon got an amoebic infection. In another study, well-meaning investigators gave iron to supplement the low levels found in Somali nomads. At the end of one month, 38 percent had infections versus 8 percent of those who had not taken the supplements.

Yet another clue: eggs are a rich source of nutrients, but their porous shells can be readily penetrated by bacteria. So how can eggs stay fresh so long? They contain lots of iron, but it is all in the yolk, none in the surrounding white. Egg white protein is 12 percent conalbumin, a molecule whose structure tightly binds iron and thereby withholds it from any bacteria that might get in. Prior to the antibiotic era, egg whites were used to treat infections.

The protein in human milk is 20 percent lactoferrin, another molecule designed to bind iron. Cow's milk has only about 2 percent lactoferrin, and breast-fed babies consequently have fewer infections than those fed from bottles. Lactoferrin is also concentrated in tears and saliva and especially at wounds, where an elevated acidity makes it especially efficient in binding iron. The researchers who discovered conalbumin predicted that there should be a similar molecule to bind iron within the body. This led to the discovery of transferrin, another protein that binds iron tightly. Transferrin releases iron only to cells that carry special recognition markers. Bacteria lack the needed code and can't get the iron. People suffering from protein deprivation may have levels of transferrin less than 10 percent of normal. If they receive iron supplements before the body has time to rebuild its supply of transferrin, free iron in the blood makes fatal infections likely—as has been a tragic outcome of some attempts to relieve victims of famine.

By now the nature of this defense is surely obvious. Iron is a crucial and scarce resource for bacteria, and their hosts have evolved a wide variety of mechanisms to keep them from getting it. In the presence of infection, the body releases a chemical called *leukocyte endogenous mediator* (LEM), which both raises body temperature and greatly decreases the availability of iron in the blood. Iron absorption by the gut is also decreased during infection. Even our food preferences change. In the midst of a bout of influenza, such iron-rich foods as ham and eggs suddenly seem disgusting; we prefer tea and toast. This is just the ticket for keeping iron away from pathogens. We tend now to think of bloodletting as an example of early medical ignorance, but perhaps, as Kluger has suggested, it did help some patients by lowering their iron levels.

It became clear in the 1970s that low iron levels associated with disease could be helpful, not harmful, but even now, Kluger and his associates find that only 11 percent of physicians and 6 percent of pharmacists know that iron supplementation may harm patients who have infections. Although the sample was small, the study illustrates the difficulty of making clinicians aware of some established scientific findings. Even top researchers may neglect to mention this adaptive mechanism. A recent study in *The New England Journal of Medicine* showed that children with cerebral malaria were more likely to recover if they were treated with a chemical that binds iron, but the article did not describe the body's natural system for binding iron during infection. The evolved mechanism that regulates iron binding is but one specific illustration of the broader principle that we should be careful to distinguish defenses from other manifestations of infection, slow to conclude that a bodily response is maladaptive, and cautious about overriding defensive responses. In short, we should respect the evolved wisdom of the body.

STRATEGIES AND COUNTERSTRATEGIES

Medical researchers are not the only ones who deal with conflicts between organisms. Ecologists and animal-behavior specialists routinely deal with predator–prey relationships, struggles between males for mating opportunities, and many other sorts of conflict. They recognize the evolutionary significance of the phenomena they observe and use such terms as *strategy* and *tactic*, *winner* and *loser*, and other indications of commitment to the adaptationist program. This approach has been richly rewarding for ecologists and others who are steeped in Darwinism. A similar approach to phenomena such as fever ought to be similarly rewarding in a field of such vital interest to all of us.

The contest between parasites and their hosts is a war, and every sign and symptom of infection can be understood in relation to the underlying strategies of one or the other belligerent. Some, like fever and iron withholding, benefit the host (defenses); others benefit the pathogen; and a few are incidental effects of the war between them. The strategies are not, of course, products of conscious thought, but they are strategies nonetheless. Bacteria that sneak into the body by pretending to be harmless are rather like Greek soldiers hiding in a wooden horse. When the manifestations of infection are related to conflicting interests, they fit neatly into categories based on their functional importance. Table 3-1 gives an overview of these categories and a guide to the organization of this chapter.

TABLE 3-1 A CLASSIFICATION OF PHENOMENA
ASSOCIATED WITH INFECTIOUS DISEASE

OBSERVATION	EXAMPLES	BENEFICIARY
Hygienic measures taken by host	Killing mosquitoes, avoiding sick neighbors, avoiding excrement	Host
Host defenses	Fever, iron withholding, sneezing, vomiting, immune response	Host
Repair of damage by host	Regeneration of tissues	Host
Compensation for damage by host	Chewing on other side to avoid tooth pain	Host
Damage to host tissues by pathogen	Tooth decay, harm to liver in hepatitis	Neither
Impairment of host by pathogen	Ineffective chewing, decreased detoxification	Neither
Evasion of host defenses by pathogen	Molecular mimicry, change in antigens	Pathogen
Attack on host defenses by pathogen	Destruction of white blood cells	Pathogen
Uptake and use of nutrients by pathogen	Growth and proliferation of trypanosomes	Pathogen
Dispersal of pathogen	Transfer of blood parasite to new host by mosquito	Pathogen
Manipulation of host by pathogen	Exaggerated sneezing or diarrhea, behavioral changes	Pathogen

How can a host guard against infection? First, it can avoid exposure to pathogens. Second, it can erect barriers to keep them out of the body and act quickly to defend and repair any breaches in the defenses. If pathogens do get beyond the outer ramparts, it can flag any cells that lack proof of identity and expel them from their entry portal. If they have breached this defense line, it can poke holes in them, poison them, starve them, do whatever is necessary to kill them. And if all this does not work, it can wall them off so that they cannot reproduce and spread. If they have done damage, it can repair it. If the damage can't be repaired immediately, it can compensate for it in some way. Some of this damage and the resulting impairment benefit neither the host nor the pathogen. They are, like the aging bomb craters on the coast of France, just incidental relics of an old battle.

The pathogens will not, of course, give up readily. Our bodies are, after all, their homes and dinners. We understandably tend to see bacteria and viruses as evils, but how anthropocentric this is! Our defenses attempt to prevent the poor streptococcus from getting even a microgram of our body tissues, but if it cannot find a way around our defenses, it will die. So, for each of our defenses, pathogens have evolved counterdefenses. They find ways to get transmitted to us and ways to breach our walls. Once inside, they hide from our sentries, attack our defenses, use our nutrients to make copies of themselves, and find ways to get those copies out of the body and to new victims, often by turning our own defenses to their own advantage. Before describing the clever stratagems used by pathogens to elude our defenses, we will discuss the defenses in more detail.

HYGIENE

The best defense is avoidance of danger; proper hygiene can prevent a pathogen from gaining that first toehold. Instinctively slapping at a mosquito is not just an attempt to spare oneself the minor annoyance of a mosquito bite. It may also prevent a long list of serious insect-borne diseases, of which malaria is the best known. Is the itch of a mosquito bite just part of the insect's nastiness? It may be merely an accidental result of the chemicals the mosquito uses to ensure that our blood flows freely, but it may also be our adaptation for avoiding future bites. Imagine what would happen to a person who did not mind being bitten by mosquitoes. And imagine how successful a mosquito could be if its biting were not noticeable!

Our tendencies to avoid contact with people who may be infectious may have the same significance. Likewise, an instinctive disgust motivates us to avoid feces, vomit, and other sources of contagion. Our tendency to defecate away from others may prevent the infection of close associates, and social pressures to conform to such practices may protect us from infection by others. The best defense against infection is avoidance of pathogens, and natural selection has shaped many mechanisms to help us keep our distance.

THE SKIN

Our skin is like the wall around an ancient city, a formidable protective barrier. It not only prevents the entry of parasites but also protects against injury by mechanical, thermal, and chemical forces. Unlike induced defenses such as fever, which are aroused only when a particular danger threatens, the skin is constantly present, always on guard. It is tough and much more resistant to puncture and abrasion than the internal tissues it protects. Minor infections here and there are harmless because the skin is constantly being sloughed off the top and renewed from below. An ink stain on the fingers will be gone in a few days, not because the ink has been absorbed or chemically altered but because the stained cells are replaced by others rising from below. Fungal growths or other potential pathogens in surface cells are constantly cast off by this rapid replacement of the epidermis. Sycamores and shagbark hickory trees seem to use the same strategy.

Not only is the skin a good defensive armor in general, it is also good in particular. Those parts of the body that are most in need of armor, such as the soles of the feet, have thicker and tougher skin right from birth. Any particular patch of skin that is subjected to repeated friction, like that at the top edge of a shoe or the tip of a cellist's finger, grows the thicker skin we call a callus. This adaptive growth, an induced defense, not only minimizes mechanical injury, it also prevents breaks in the skin that could provide entrances for pathogens.

Some of our most useful hygienic behaviors help maintain the skin's barrier. The most obvious are behaviors that keep nasty things off the skin. Scratching and other grooming maneuvers remove external parasites, important sources of discomfort and disease transmission for most people during most of human history and still problems in less fortunate societies. Benjamin Hart, a veterinarian from the University of California at Davis, has shown just how crucial grooming is to preventing illness in animals. An animal that cannot groom is quickly infested with fleas, ticks, lice and mites, and will lose weight and fall ill. The mutual grooming of monkeys is not just a ritual, it is preventive health care.

PAIN AND MALAISE

J ust as an itch can motivate defensive scratching, pain is an adaptation that can lead to escape and avoidance. The skin, sensibly enough, is highly sensitive to pain. If it is being damaged, something is clearly wrong, and all other activities should be dropped until the damage is stopped and repair can begin. Other kinds of pain can also be helpful. While an abstract realization that chewing is impaired because of an abscessed tooth might possibly lead to more chewing with other, unimpaired teeth, the tormenting pain of a toothache far more effectively prevents the pressure on the tooth that would delay healing and spread bacteria. Continued pain from infection or injury is adaptive because continued use of damaged tissue may compromise the effectiveness of other adaptations, such as tissue reconstruction and antibody attacks on bacteria. Pain motivates us to escape quickly when our bodies are being damaged, and the memory of the pain teaches us to avoid the same situation in the future.

The simplest way to determine the function of an organ like the thyroid gland is to take it out and then see how the organism malfunctions. The capacity for pain cannot be removed, but very occasionally someone is born without it. Such a pain-free life might seem fortunate, but it is not. People who cannot feel pain don't experience discomfort from staying in the same position for long periods, and the resulting lack of fidgeting impairs the blood supply to the joints, which then deteriorate by adolescence. People who cannot feel pain are nearly all dead by age thirty.

Generalized aches and pains, or merely feeling out of sorts (malaise, in medical terminology), are also adaptive. They encourage a general inactivity, not just disuse of damaged parts. That this is adaptive is widely recognized in the belief that it is wise to stay in bed when you are sick. Inactivity also likely favors the effectiveness of immunological defenses, repair of damaged tissues, and other host adaptations. Medication that merely makes a sick person feel less sick will interfere with these benefits. This is fine when patients are well informed about the risks and realize that they are sicker than they feel and should make a special effort to take it easy. Otherwise, a drug-induced feeling of well-being may lead to activity levels that interfere with defensive adaptations or repairs.

DEFENSES BASED ON EXPULSION

The body must have openings for breathing, for the intake of nutrients and expulsion of wastes, and for reproduction. Each of these openings offers pathogens an invasion route, and each is endowed with special defense mechanisms. The constant washing of the mouth with saliva kills some pathogens and dislodges others so they can be destroyed by the acid and enzymes in the stomach. The eyes are washed by tears laden with defensive chemicals and the respiratory system by antibody and enzyme-rich secretions that are steadily propelled up to the throat, where they can be swallowed so the invaders can be killed and the protein in the mucus recycled. The ears secrete an antibacterial wax. Projections inside the nose, called turbinates, provide a large surface that warms, moistens, and filters pathogens from the incoming air. Mouth-breathers don't get the full benefit of this defense and are more subject to infection. The nose and ears have hairs strategically arrayed to keep out insects.

The defenses at each body opening can be quickly increased if danger threatens. Irritation of the nose by a viral infection provokes the discharge of such copious mucus that one can go through a whole box of tissues in a day. Millions of people use nasal sprays each year to block this useful response, but there are remarkably few studies that have investigated whether the use of such devices delays recovery from a cold. If they do not demonstrably delay recovery, as seems to be the case from the limited data, it would be evidence that a runny nose is not a defense but an example of a pathogen manipulating the host's physiology in order to spread itself. Sneezing is obviously a defensive adaptation, but not every sneeze need be adaptive for the sneezer. Some sneezing may possibly be an adaptation that viruses use to disperse themselves.

Irritation deeper in the respiratory tract induces coughing. Coughing is made possible by an elaborate mechanism that involves detecting foreign matter, processing this information in the brain, stimulating a cough center at the base of the brain, and then coordinating muscle contractions in the chest, the diaphragm, and the tubes in the respiratory tract. All along the lining of these tubes tiny hairs called cilia beat in a steady rhythm, sweeping pathogen-trapping mucus upward. In the urinary tract, periodic flushing washes pathogens away along with the cells on the surface of the urethral lining, which are systematically shed like those on the skin. When the bladder or urethra becomes infected, urination understandably becomes more frequent.

The digestive system has its own special defenses. Bacterial decomposition and fungal growths produce repulsive odors, the repulsiveness being our adaptation to be disinclined to put bad-smelling things into our mouths. If something already in the mouth tastes bad, we spit it out. Taste receptors detect bitter substances that are likely to be poisonous. After we swallow something, there are receptors in the stomach to detect poisons, especially those made by bacteria that multiply in the gastrointestinal tract. When absorbed toxins enter the circulation, they pass by a special group of cells in the brain, the only brain cells directly exposed to the blood. When these cells detect toxins, they stimulate the brain's chemoreceptor trigger zone to respond first with nausea and then with vomiting. This is why so many drugs are so nauseating, especially the toxic ones used for cancer chemotherapy.

Circulating toxins almost always originate in the stomach, so it is easy to see how vomiting is useful: it ejects the toxin before more is absorbed. What about nausea? The distress of nausea discourages us from eating more of the noxious substance, and its memory discourages future sampling of whatever food seemed to cause it. Just a single experience of nausea and vomiting after eating a novel food will cause rats to avoid it for months; people may avoid it for years. This remarkably strong onetime learning was named the "sauce béarnaise syndrome" by Martin Seligman, a psychologist who recognized its significance after contemplating the untimely loss of his gourmet dinner. Why is the body capable of such a strong association after a single exposure to a food that produces illness? Imagine, for a moment, what would happen to the person who ate poisonous foods repeatedly.

The other end of the intestinal tract has its own defense, diarrhea. People understandably want to stop diarrhea, but if relief comes from merely blocking the defense, there is likely to be some penalty. Indeed, H. L. DuPont and Richard Hornick, infectious disease experts at the University of Texas, found just this. They infected twenty-five volunteers with *Shigella*, a bacterium that induces severe diarrhea. Those who were treated with drugs to stop the diarrhea stayed feverish and toxic twice as long as those who did not. Five out of six who received the antidiarrheal drug Lomotil continued to have *Shigella* in their stools, compared to two out of six who did not receive the drug. The researchers concluded, "Lomotil may be contraindicated in shigellosis. Diarrhea may represent a defense mechanism." Consumers will no doubt want to know when they should and should not take such medications for more commonplace diarrhea, but the needed research has not been done. There are dozens of studies of side effects, of safety, and of the effectiveness of medications that block diarrhea, but few consider the consequences of the main effect of blocking a normal defense.

Our reproductive machinery requires yet another opening, which in males is the same as that of the urinary tract, whose defenses thereby do double duty. Women have a separate opening that poses a special problem for defense against infection. While the female reproductive tract uses many defenses, such as cervical mucus and its antibacterial properties, one largely unappreciated defense is the normal outward movement of secretions that makes it difficult for bacteria and viruses to gain access. These secretions move steadily from the abdominal cavity through the fallopian tubes, uterus, cervix, and vagina to the outside. There is one noteworthy exception to this constant downstream movement. Sperm cells swim upstream, from the vagina through the uterus into the fallopian tubes and the pelvic cavity. Unusually small for human cells, sperm are still large compared to bacteria. Potential pathogens can stick to sperm cells and be transported from the outside to deep within a woman's reproductive system.

Only recently has the threat of sperm-borne pathogens been recognized. Biologist Margie Profet notes that menstruation has substantial costs and argues that it must therefore give some compensating benefit. After a consideration of the evidence, she concluded that many aspects of menstruation seem designed as an effective defense against uterine infection. The same anti-infection benefits that come from sloughing off skin cells are achieved by the periodic extrusion of the lining of the uterus. This is supported by evidence that menstrual blood differs from circulating blood in ways that make it more effective in destroying pathogens while minimizing losses of nutrients. Studies of menstruation in other mammals suggest that each species menstruates to just the extent appropriate for its vulnerability to sperm-borne pathogens. The threat is small for species that restrict their sexual behavior to widely separated fertile periods, but women's continuous sexual attractiveness and receptivity are largely unrelated to the ovulatory cycle. This extraordinary amount of human sexual activity may have its benefits, as we will discuss in Chapter 13, but it substantially increases the risk of infection. This risk may be responsible for the unusually profuse human menstrual discharge, as compared to other mammals'.

We have mentioned several times that evolutionary hypotheses need to be and can be tested. Beverly Strassmann has mounted a challenge to the hypothesis that menstruation protects against infection. She maintains that the pathogen load in the reproductive tract is the same before and after menstruation, that menstruation does not increase when there is infection, and that there is no consistent relationship between the amount of sperm females in a particular species are exposed to and the amount of menstrual flow. As an alternative explanation, Strassmann proposes that the degree of shedding or reabsorption of the uterine lining depends on the metabolic costs of maintaining it or shedding it, a hypothesis that she supports with comparisons between species and the relationship between menstruation and the body weight of the female and her neonate. Obviously, we have not heard the last word on this issue.

MECHANISMS TO ATTACK INVADERS

Vertebrates in general, and mammals in particular, have amazingly effective immunological defenses that are in essence a system of carefully targeted chemical warfare. Cells called macrophages constantly wander the body searching for any foreign protein, whether from a bacterium, a bit of dirt in the skin, or a cancer cell. When they find such an intruder, the macrophages transfer it to a helper T cell, which then finds and stimulates whichever white blood cells can make a protein (called an *antibody*) that binds specifically to that particular foreign protein (an *antigen*). Antibodies bind to antigens on the surfaces of bacteria, thereby impairing the bacteria and also labeling them for attack by specialized larger cells. If the antigens persist, say during a continuing bacterial infection, they stimulate the production of ever more of the cells that make that specific antibody, so that the bacteria are destroyed at an ever-increasing rate. Whatever is recognized as a properly functioning part of the body is permitted to remain. All else—disease organisms, cancerous tissue, organs transplanted from other individuals—is attacked.

How does the body recognize cells as its own? Each cell has a molecular pattern on its surface, called the *major histocompatibility complex* (MHC), which is like a photo ID card. Cells that have a valid MHC are left alone, but those that have a foreign or missing MHC are attacked. Interestingly, when cells are infected, they transport protein from the invader to the MHC, where it is bound. Like individuals with obviously fake ID cards, such cells are priority targets for the killer cells of the immune system. The adenovirus, a common cause of sore throats, has found a way to get around this defense. It makes a protein that blocks the ability of the cell to move foreign proteins to the MHC. In essence, it prevents the infected cell from signaling that it has been invaded.

The operation of the MHC system is a vivid example of altruism in its biological sense. An infected cell "volunteers" for destruction for the good of the rest of the body. This is like a soldier with plague asking his comrades to destroy him before he infects them. The analogy, however, is false in one crucial respect. The cell's comrades are genetically identical, and its only chance for passing on its genes lies in the success of the whole organism. Soldiers, however, seldom share foxholes with identical twins and are understandably less likely to volunteer for elimination.

The weapons of the immune system are truly fearsome. They include general inflammation, several kinds of antibodies—each specialized for a different group of opponents—and a series of chemicals (the complement system), five of which attack the targeted cells, boring holes in their membranes and digesting them. Despite these weapons, some invaders can nonetheless persist. When a clump of bacteria can be neither expelled nor destroyed, it may be walled off by a membrane that keeps it away from vulnerable tissues. The tubercles from which tuberculosis gets its name are the best-known example, but analogous imprisonment of roundworms and other multicellular parasites has also been important throughout most of human evolution.

DAMAGE AND REPAIR

In the contest with their host, pathogens must rob the host to secure their own nourishment. Various bacteria and the protozoa that causes amoebic dysentery secrete enzymes that digest nearby host tissues and then absorb the products of digestion. Others literally eat through host tissues, for example, filaria worms, which live in the anterior part of the eye, or the larvae of another species of worm, *Angiostrongylus cantonensis*, which burrow through the brain. Both of these defend themselves with secretions that inhibit inflammation. Still others, such as the trypanosomes, a group of protozoans that cause diseases such as African sleeping sickness, live in the bloodstream and absorb nutrients directly from the plasma. Whatever the means, parasites secure their resources from the host and then use them for their own maintenance, growth, and reproduction.

These activities of pathogens incidentally damage the host, but this damage is not a pathogen adaptation. It does not do a tapeworm any good to have its host malnourished. It does not do the malarial parasite any good to destroy its host's blood cells (unless, perhaps, this frees up iron for use by the parasite). Most often, the opposite must be true. The survival and well-being of the parasite depend on the host's continued survival and ability to provide it with nourishment and shelter. Such incidental damage must therefore be considered a cost to both host and pathogen.

The cost may be a general reduction in host resources or an obviously localized destruction. Bacteria that attack bone where a tooth is rooted cause structural damage and perhaps the loss of the tooth. The bacteria that cause gonorrhea may erode the connective tissue and cartilage of joints, causing functional impairment. Hepatitis viruses may destroy substantial portions of the liver, so that all liver functions, such as the clearing of toxins from the blood, become less effective. Such functional impairments are simply incidental consequences of pathogen adaptations. It does not do bacteria any good to make the host's chewing less effective or its running less rapid.

It's important to keep damage conceptually separate from any resulting functional impairment. The damage causes the impairment, which can then itself be a cause for another host adaptation, which we call *compensatory adjustment*. There are many examples, some much more subtle than chewing on the left side of your mouth if it hurts to chew on the right. For instance, when disease-damaged lungs become less effective at oxygenating the blood, this may be partly compensated for by an increase in blood hemoglobin concentration. The body has a mechanism that monitors the oxygen level in the blood. If there is too little, whether from living at a high altitude or from lung damage, the body makes more erythropoietin, a hormone that stimulates the production of more red blood cells.

Another obvious host adaptation is repair of damage. Natural selection has adjusted the ability to regenerate various tissues according to how useful it would normally be to do so. The skin, which is often damaged, is a first line of defense against pathogens and injuries. As might be expected, it quickly regenerates and rapidly recovers its protective capabilities. Other structures that regenerate quickly are the lining of the gut and organs such as the liver, which are in open communication with the gut and therefore with the outside world and its infectious agents. By contrast, the heart and especially the brain are less accessible to most pathogens. If pathogens do gain access and cause serious damage, it is ordinarily fatal, so regenerative capabilities would rarely be of benefit.

PATHOGEN EVASION OF HOST DEFENSES

So far we have mentioned only one kind of pathogen adaptation, the ability to nourish itself in the body of the host. We can also expect it to have evolved ways of shielding itself from the host's efforts to destroy, expel, or sequester it. We will now turn to one such mechanism, *evasion of host defenses*.

The first trick for many parasites, once inside the body, is to gain entrance to cells. Invaders may accomplish this just as door-to-door peddlers do, by appearing to offer something else. The rabies virus binds to acetylcholine receptors as if it were a useful neurotransmitter; the cowpox virus to epidermal growth-factor receptors as if it were a hormone; and the Epstein-Barr virus (which causes mononucleosis) to a C4 receptor. Rhinovirus, a common cause of colds, binds to the intercellular adhesion molecule (ICAM) on the surface of the lymphocytes that line the respiratory tract. This is extremely clever, since attacking lymphocytes releases chemicals that greatly increase the number of ICAM binding sites, thus providing many more openings by which the virus can enter cells.

Another trick is to evade the immune system. The trypanosome that causes African sleeping sickness does this by rapidly changing its disguises. It takes the body about ten days to make enough antibodies to control the trypanosome, but on about the ninth day, the trypanosome changes its disguise by exposing an entirely new surface layer of proteins, thus escaping attack by the antibodies. The trypanosome has genes for more than a thousand different antigenic coats and so can live on for years in the human host, always one step ahead of the immune system. Two other common bacteria use similar strategies. *Hemophilus influenza,* a common cause of meningitis and ear infections, and *Neisseria gonorrhoeae,* the cause of gonorrhea, both have what seem to be flaws in the genetic mechanisms that make their surface proteins. The seeming errors are useful, however, because the resulting variation makes it hard for our immune systems to keep up with the random changes.

Malarial parasites have special surface proteins that allow them to bind to the walls of blood vessels so that they are not swept to the spleen, where they would be filtered out and killed. The genes that code for these binding proteins in malarial parasites mutate at a rate of 2 percent per generation, just enough so that the immune system cannot lock in on the organism. The pneumococcal bacteria that cause pneumonia use a different trick to circumvent the immune system. They have "slippery" polysaccharides on their surface that white blood cells can't get a grip on. The body copes with this by making chemicals called opsonins, which bind to the microbe like handles that the antibodies can grab.

Another common evasion is a chemical analog of a disguise a spy might use behind enemy lines. The external chemistry of some bacteria and some worms is so similar to that of human cells that the host may have difficulty in recognizing them as foreign. (Thus antibodies sometimes attack both invader and host cells.) The streptococcus bacterium, a longtime associate of humans, is especially adept at this trick. The antibodies to some strains cause rheumatic fever, in which a person's antibodies attack his or her own joints and heart. Similar antibody attack on nerve cells in the basal ganglia of the brain can cause Sydenham's chorea, with its characteristic uncontrollable muscle twitches. Interestingly, many patients who have obsessive-compulsive disorder, a psychiatric illness characterized by excessive hand washing and fear of accidentally harming others, had Sydenham's chorea in childhood. There is now growing evidence that the brain areas involved in obsessive-compulsive disorder are very close to those damaged by Sydenham's chorea. Thus, some cases of obsessive-compulsive disorder may result from the arms race between the streptococcus and the immune system.

Chlamydia, today's most common cause of venereal disease, does the equivalent of hiding in the police station. It enters white blood cells and then builds a wall to prevent itself from being digested. Schistosomes of the *mansoni* type go a step further and essentially steal police uniforms. These parasites, a serious cause of liver disease in Asia, pick up blood-group antigens so that they may look to the immune system like our own normal blood cells.

ATTACK ON HOST DEFENSES

Pathogens not only attempt to shield themselves from the weaponry of the host, they also have destructive weaponry of their own. The bacterium that causes most simple skin infections, *Staphylococcus aureus*, secretes a neuropeptide that blocks the action of Hageman's factor, a crucial first step in useful inflammation. Bacteria that cannot secrete this peptide do not cause infection. Even the common streptococcal bacteria that cause so many sore throats make streptolysin-O, which kills white blood cells. Vaccinia, the virus that causes cowpox, makes a protein that inhibits the complement system, an important host defense, as noted previously. Why doesn't the complement system attack our own cells? In part because our cells have a layer of sialic acid, a chemical that protects them from attack by the complement system. Sure enough, certain bacteria, in this case the K1 strain of the common *E. coli* that live in our guts, are able to cover themselves in sialic acid and thus gain protection from the complement system.

One of the great dangers of serious infection with certain kinds of bacteria is shock, a decrease in blood pressure that can be rapidly fatal. Shock is caused by chemical lipopolysaccharide (LPS) formed by the bacteria. Superficially, it would seem that LPS is a toxin made by bacteria to harm us, but, as researcher Edmund LeGrand has noted, this is unlikely, because LPS is a necessary component of the cell wall of this whole group of bacteria. Hosts recognize this reliable cue to the presence of dangerous infection and react strongly—sometimes too strongly. Here is an example of a defensive weapon that can turn on its bearer.

The human immunodeficiency virus (HIV), the virus that causes AIDS, hides in the helper T cells that bring antigens to the attention of the immune system. These cells have a protein in their outer membrane called CD-4, to which the HIV binds to gain entrance to cells. This protein on HIV would make it vulnerable to the immune system, except that it is hidden in deep crevices in the viral wall. As HIV kills helper T cells, it incidentally causes the victim to be ever more vulnerable to other infections and cancer, the problems that eventually kill a person who has AIDS.

OTHER PATHOGEN ADAPTATIONS

There remain two related categories of parasite adaptation. No matter how well a pathogen survives and proliferates in a host, it must have a dispersal mechanism so that it can get itself or its descendants into other hosts. For external parasites this can be rather easy. Lice and the fungus that causes ringworm, for example, are readily spread by personal contact. Internal parasites face greater problems. Those that can regularly get onto the skin have the possibility of contact with other susceptible individuals. Cold viruses and intestinal bacteria may get onto hands or other surfaces and be spread by handshakes or more intimate contact.

Microorganisms in the bloodstream are not likely to be spread in this way. Many can be transmitted only with the help of biting insects or other transport agents (vectors). Malaria is a well-known example. If there are about ten malarial parasites in the dispersal stage (called · gametocytes) in each milligram of blood and a mosquito sucks up three milligrams, it will be taking in about thirty gametocytes. The next item on the mosquito's agenda is to convert this rich blood meal into eggs and get them fertilized and laid in an environment suitable for development. Meanwhile, the sexually produced offspring of the malarial plasmodia have migrated to the mosquito's salivary glands, where they transform into an infectious stage in the fluid that will be used to inhibit clotting when the mosquito sucks up its next blood meal. The mosquito then unwittingly injects the plasmodia into the next victim. An enormous variety of insects and other organisms can serve as vectors of human diseases.

Another kind of parasitic adaptation is technically termed *host manipulation*. By subtle chemical influence a parasite may gain some control over the machinery of the host's body and cause that machinery to serve the interests of parasite rather than host. Many curious examples are known from many groups of organisms. The tobacco mosaic virus causes its host to enlarge the pores between adjacent tobacco cells enough to allow the virus particles to pass through and infect other cells. One kind of parasitic worm alternates its life stages between ants and sheep, just as malarial parasites must alternate between vertebrate hosts and mosquitoes. The worm is effectively transmitted from an ant to a sheep because it enters certain sites in the ant's nervous system where it causes the ant to climb to the top of a blade of grass and hang on, unable to let go. This greatly increases the likelihood that the ant will be eaten by a sheep. Another kind of worm alternates between snails and gulls. It causes the snail, which is ordinarily hard to find in the tangled growths of shallow coastal waters, to crawl up to a high level of bare rock or sand and stay there. It is then easily seen and eaten by a gull.

The rabies virus offers a particularly remarkable and gruesome example of how a pathogen can manipulate a host's behavior. After gaining entrance to the body, usually via the bite of an infected individual, the rabies virus moves along nerve fibers to the brain, where it concentrates in regions that regulate aggression. It can then make the host attack and bite, thereby infecting other individuals. It also paralyzes the victim's swallowing muscles, thus causing virus-laden saliva to build up in the mouth, increasing the likelihood of transmission and incidentally causing the victim to have the terror of choking on fluids that originally gave the disease the name hydrophobia.

Perhaps the most important human examples of manipulation by pathogens are the sneezing, coughing, vomiting, and diarrhea triggered by bacteria and viruses. At some stage in the history of an infection, this expulsion will serve the interests of both host and microbe. The host is benefited by having fewer pathogens attacking its tissues, the microbe by an increased chance of finding other hosts. The losers in this game are currently healthy but vulnerable individuals. A chemical released by cholera bacteria reduces absorption of liquid from the bowel, causing profuse diarrhea that, in a society without well-developed public hygiene, can effectively spread an epidemic.

Sometimes we are successfully manipulated by our parasites, at other times we successfully resist manipulation, and in still other situations there is some intermediate resolution. Any given example of such a conflict is likely to be at an evolutionary equilibrium and have a consistent outcome. Conflicts are often decided in favor of the antagonist that has the most to gain from winning. If someone is sneezing twice as often as would be ideal for the control of a cold virus, that is not likely to be a great burden of lost time or energy, but it may nearly double the rate at which the virus reaches new hosts. This is just the sort of contest we would expect the virus to win. How frequently are expulsion mechanisms exaggerated by pathogens beyond what would be optimal to a human host? The paucity of evidence on this issue shows the habitual neglect of such evolutionary questions.

A FUNCTIONAL APPROACH TO DISEASE

We end this chapter by making three remarks about Table 3-1 (page 32), which classifies the signs and symptoms of infectious disease according to their functions. First, a functional classification of the signs and symptoms of disease is important and useful. In order to choose appropriate treatment, we need to know if the cough, or other symptom, benefits the patient or the pathogen. We also need to know if the pathogen is manipulating the host or attacking its defenses. Instead of just relieving symptoms and trying, perhaps ineffectively, to kill the pathogen, we can analyze its strategies, try to oppose each of them, and try to assist the host in its efforts to overcome the pathogen and repair the damage. The second point is that the classification is really rather simple and obvious.

Now for the third point: When and by whom do you think the ideas in this chapter were first proposed? Was it by some nineteenth-century medical researcher building on the ideas of Pasteur and Darwin as well as the rapidly expanding body of knowledge of parasite life histories? No. The classification scheme used in our table and throughout this chapter was first proposed at the University of Michigan in 1980 by Paul Ewald, an ornithologist and evolutionary biologist now at Amherst College. And when did the ideas in this chapter first become standard elements in the thinking of physicians and medical researchers? The answer to this question is a simple and discouraging *not yet*. We do not mean that physicians never intuitively think in the categories formalized by Ewald. We merely mean that they have not been explicitly taught to use them and that deficiencies of training make it easy to neglect these essential ideas in thinking about infectious disease. There is hope, which is especially evident in the proceedings from several recent conferences that have emphasized the benefits of interchange between evolutionists and infectious disease experts. But it will still be years before this sort of material becomes part of the regular medical curriculum.

Why has the medical profession not taken advantage of the help available from evolutionary biology, a well-developed branch of science with great potential for providing medical insights? One reason is surely the pervasive neglect of this branch of science at all educational levels. Religious and other sorts of opposition have minimized the impact in general education of Darwin's contributions to our understanding of ourselves and the world we live in. There has also been a peculiar neglect of evolution in the training of physicians and medical researchers, a matter discussed further in Chapter 15.

Still another reason is that many of the evolutionary ideas of greatest bearing on medicine have only been formulated in recent years. These ideas are often simple and not very different from common sense—once they are pointed out. Yet their recognition and the appreciation of their importance have come only in the past few years, far behind the development and application of many really complex and subtle branches of physical science and molecular biology. Exactly why the application of evolutionary biology to medicine and other aspects of human life has advanced so slowly after its magnificent inception in 1859 is a question that ought to be getting major attention from historians of science.

4

AN ARMS RACE
WITHOUT END

E very time a nation or a tribe designs a new weapon, a competing nation or tribe will soon devise a counterweapon.
Thus spears and swords gave rise to shields and body
armor, and radar defenses to the Stealth Bomber. Likewise,
the evolutionary origin of a predator's improved hunting technique
can be countered by the prey's improved armor, evasive tactics, or
other defensive adaptation, which is then met by countermeasures
from its predators. If foxes start running faster, rabbits are selected to
run even faster so that foxes must run faster still. If foxes' eyesight
improves, this selects for rabbits that blend better with the background, which may select for foxes that can locate rabbits by smell,
which in turn may select for rabbits that tend to move downwind
from foxes. Thus predator and prey coevolve in an escalating cycle of
complexity. Biologists have named this idea the *Red Queen Principle*
after Lewis Carroll's Red Queen, who explained to Alice, "Now,
here, you see, it takes all the running you can do, just to keep in the
same place."

Like contests between predators and prey, wars between hosts
and parasites initiate escalating arms races that require extravagant,
harmful expenditures and create extraordinarily complex weapons
and defenses. Just as political powers sometimes put more and more
of their energies into weaponry and defense to keep from being

dominated by opponents, hosts and parasites must both evolve as fast as they can to maintain their current levels of adaptation. There comes a point where the expense of an arms race is so great that the organism, political or biological, is hard put to meet other basic needs, but the cost of losing it is so great that enormous expenses may nonetheless be maintained. We are in a relentless all-out struggle with our pathogens, and no agreeable accommodation can ever be reached.

The relationships between hosts and parasites are so competitive, wasteful, and ruthlessly destructive that arms-race terminology offers an entirely appropriate framework for describing them. The rest of this chapter explains this point of view, but for an introduction, just try to imagine the magnitude of the personal tragedy that infectious agents have caused throughout human history, until just a few decades ago. The mother of one of the authors (Williams) was orphaned at age nine by meningitis. He has a sister whose best friend died suddenly of acute appendicitis in fourth grade. Our microscopic enemies take no account of individual merit or importance. Shortly before Calvin Coolidge succeeded to the presidency of the United States, his sixteen-year-old son got a blister on his foot while playing tennis but bravely went on playing. The blister broke open and became infected, and in two weeks the boy was dead. As a result, the president of the United States was an ineffective emotional cripple (as even his admirers concede) throughout the ensuing campaign and his one term in office.

The analogy between international arms races and host-parasite coevolution is not exact. The Pentagon can plan new weapons on the drawing board and then try out models and prototypes. It has the benefit of rational planning, fresh starts, and trial-and-error tinkering. In evolution, there are no think tanks systematically devising ways of putting scientific knowledge to new destructive or defensive uses. No plans contribute to evolution, and there can be no fresh starts. *Evolution consists entirely of trial-and-error tinkering.* The slightly different variants of every generation compete in the game of life. Some achieve a higher reproductive output than others, and the population averages shift slightly in their direction. The process is slow and unguided—in some ways misguided—but there is no limit to the precision and complexity of adaptation that the Darwinian process can generate.

PAST VERSUS CURRENT EVOLUTION

Many microbiologists incorrectly assume that hosts and their pathogens are usually in a state of slow evolutionary change toward some optimal future state, usually of active cooperation. This is a grossly unrealistic idea. Both pathogens and hosts must normally maintain close-to-stable equilibria by making trade-offs between competing values, such as growth rates and defensive activities. At equilibrium, a unit of improvement of one adaptation would require more than one unit of loss of another. A leaner rabbit might run faster, but at some point the benefit of still greater speed would not be worth the added risk of starvation. Likewise, our fever response is presumably optimized, at least for historically normal conditions. Higher and more frequent fever would make us less vulnerable to pathogens but would be more than counterbalanced by the costs of tissue damage and nutrient depletion. This will be true as long as the environment stays constant. If circumstances change, some of the optima for both host and pathogen will likely change. If bacterial pathogens are artificially kept in check for many generations, this may select for a decreased fever response, but if our technology fails and we become vulnerable again, we might recover a heightened fever response.

In all of this book's other chapters we deal mainly with features of human biology established by long-term historical processes. In the present chapter we will discuss evolutionary changes that can occur within the next year, or perhaps maybe even next week. Because pathogens reproduce so rapidly, they also evolve rapidly.

Some of our defenses against disease, such as sickle cell hemoglobin, have evolved markedly in the last ten thousand years, during which we have had perhaps three hundred generations. The species as a whole has evolved significantly higher resistance to a few epidemic diseases such as smallpox and tuberculosis in the last few centuries, perhaps a dozen generations. Compare this to a bacterium's three hundred generations in a week or two and the even faster reproduction of a virus. Bacteria can evolve as much in a day as we can in a thousand years, and this gives us a grossly unfair handicap in the arms race. We cannot evolve fast enough to escape from microorganisms. Instead, an individual must counter a pathogen's evolutionary changes by altering the ratios of its various kinds of antibody-producing cells. Fortunately, the number and diversity of these chemical weapons factories are enormous and at least partly compensate for our pathogens' great evolutionary advantage.

From an immunological perspective, an epidemic may change a human population dramatically. Those individuals who have contracted a disease and recovered will likely be immune to reinfection because they harbor vastly increased concentrations of the lymphocytes that make the antibodies that are most destructive of that particular pathogen. Adult immunity to childhood diseases such as mumps depends not on changing human gene pools but on changing the concentrations of different kinds of antibodies within each individual.

Small size gives our pathogens another advantage: their enormous numbers. Each of us carries around (mostly in our digestive and respiratory systems) more bacterial cells than there are people on Earth. These enormous numbers mean that even improbable sorts of mutations will occur with appreciable frequency and that any mutant bacterial strain with even the most minute advantage over the others will soon prevail numerically. We can expect our pathogens' quantitative characteristics to evolve rapidly to whatever values are optimal for present circumstances.

In some catastrophic epidemics, a human population can evolve a higher level of resistance to an infectious disease in mere months. When Europeans first arrived in the New World, for example, some European diseases quickly killed as much as 90 percent of a Native American community in a short time. If the Native Americans' vulnerability had had any genetic basis, the genes of the lucky few who survived the epidemic would have become proportionately more frequent, and we could say that the population, in this limited sense, evolved a higher resistance. This is an extreme example. More often, a human gene pool will be little changed by an epidemic, while the pathogen's features may evolve dramatically.

BACTERIAL RESISTANCE TO ANTIBIOTICS

Perhaps the greatest medical advance of this century, and one of the greatest of all time, was the discovery that toxins produced by fungi could kill the bacteria that cause human disease. While arsenic compounds had been used for syphilis since Paul Ehrlich introduced them in 1910, the antibiotic era did not really begin until Alexander Fleming noted one day in 1929 that bacteria in his petri dishes would not grow properly in the vicinity of contaminating colonies of the mold *Penicillium*. Why should this have been? Why did the most effective antibiotics come from molds? Antibiotics are chemical warfare agents that evolved in fungi and bacteria to protect them from pathogens and competitors. They were shaped by millions of years of trial-and-error selection to exploit the special vulnerabilities of bacteria but to be nontoxic to the fungi.

A wide variety of fungal and bacterial products that are safe for most people can devastate the bacteria that cause tuberculosis, pneumonia, and many other infections. For several decades now, these antibiotics have given economically advanced societies a golden age of relief from bacterial disease. A combination of public health measures and antibiotics made the death rates from infectious disease fall so rapidly that in 1969 the Surgeon General of the United States felt justified in announcing that it was "time to close the book on infectious disease."

Like other golden ages, this one may be short-lived. Dangerous bacteria, most notably those that cause tuberculosis and gonorrhea, are now more difficult to control with antibiotics than they were ten or twenty years ago. Bacteria have been evolving defenses against antibiotics just as surely as they have been evolving defenses against our natural weaponry and that of fungi throughout their evolutionary history. As Mitchell Cohen of the Centers for Disease Control and Prevention put it recently, "Such issues have raised the concern that we may be approaching the post-antimicrobial era."

Indeed we may. Consider staphylococcal bacteria, the most common cause of wound infection. In 1941, all such bacteria were vulnerable to penicillin. By 1944, some strains had already evolved to make enzymes that could break down penicillin. Today, 95 percent of staphylococcus strains show some resistance to penicillin. In the 1950s, an artificial penicillin, methicillin, was developed that could kill these organisms, but the bacteria soon evolved ways around this as well, and still new drugs needed to be produced. The drug ciprofloxacin raised great hopes when it was introduced in the mid-1980s, but 80 percent of staphylococcus strains in New York City are now resistant to it. In an Oregon Veterans' Administration hospital, the rate of resistance went from less than 5 percent to over 80 percent in a single year.

In the 1960s, most cases of gonorrhea were easy to control with penicillin, and even the resistant strains responded to ampicillin. Now 75 percent of gonococcal strains make enzymes that inactivate ampicillin. Some of these changes were apparently a result of standard chromosomal mutation and selection, but bacteria have another evolutionary trick. They are themselves infected by tiny rings of DNA called plasmids, which occasionally leave a part of their DNA behind as a new part of the bacterial genome. In 1976, it was discovered that the bacteria that cause gonorrhea had gotten the genes that code for penicillin-destroying enzymes via plasmids from *Escherichia coli*, bacteria that normally live in the human gut, so that now 90 percent of the gonorrheal bacteria in Thailand and the Philippines have become resistant. Similarly, the gene that caused antibiotic resistance in a strain of *Salmonella flexneri* that caused a 1983 outbreak of severe diarrhea on a Hopi Indian reservation was traced back to a woman who had been taking long-term antibiotics to suppress an *E. coli* urinary tract infection.

The list of threats we face from antibiotic-resistant bacteria is long and frightening. A plasmid-mediated ability to prevent binding of erythromycin has made over 20 percent of pneumococcal bacteria resistant to treatment with that drug in France. Some strains of the cholera now threatening thousands in South America are resistant to all five previously effective drugs. Amoxicillin is no longer effective against 30 to 50 percent of pathogenic *E. coli*. It appears that we are indeed running, together with the Red Queen, as fast as we can just to stay in the same place.

Perhaps most frightening of all, one third of all cases of tuberculosis in New York City are caused by tuberculosis bacilli resistant to one antibiotic, while 3 percent of new cases and 7 percent of recurrent cases are resistant to two or more antibiotics. People with tuberculosis resistant to multiple drugs have about a 50 percent chance of survival. This is about the same as before antibiotics were invented! Tuberculosis is still the most common cause of death from infection in developing countries, causing 26 percent of avoidable adult deaths and 6.7 percent of all deaths. TB rates in the United States fell steadily until 1985 but have increased 18 percent since then. About half of these cases resulted from impaired immune function in people with AIDS, the rest from increased opportunity for contagion and drug-resistant pathogens.

Increasing tolerance to antibiotics is the most widely known and appreciated kind of pathogen evolution. Since their discovery in the

1950s, an enormous number of studies have established many medically important conclusions:

1. Bacterial resistance to antibiotics arises not by the gradual development of tolerance by individual bacteria but by rare gene mutations or new genes introduced by plasmids.
2. Gene mutations can be transmitted by plasmid infection or other processes to different species of bacteria.
3. The presence of an antibiotic causes the initially rare mutant strain to increase and gradually replace the ancestral type.
4. If the antibiotic is removed, ancestral strains slowly replace the resistant forms.
5. Mutations within a resistant strain can confer still greater resistance, so that increasing the dose of an antibiotic may be effective only temporarily.
6. Low concentrations of an antibiotic, which may retard bacterial growth only slightly, will eventually select for strains that resist the slight retardation.
7. Mutations that confer still higher levels of resistance arise in such partially adapted strains more often than in the original nonresistant strain.
8. Resistance to one antibiotic may confer resistance to another, especially if the two are chemically related.
9. Finally, the disadvantage of resistant strains in the absence of an antibiotic is gradually lost by further evolutionary changes, so that resistance can prevail even where no antibiotics have been used for a long time.

The implications of these findings for medical practice are now widely appreciated. If one antibiotic doesn't alleviate your disease, it may be better to try another, instead of increasing the dose of the first. Avoid long-term exposure to antibiotics; taking a daily penicillin pill to ward off infection is accepted therapy for some conditions, such as infection of vulnerable heart valves, but has the incidental effect of selecting for resistant strains. Unfortunately, we may often be exposed to this side effect without knowing it, by consuming meat or eggs or milk from animals routinely dosed with antibiotics. This is a hazard that has recently provoked conflict between food producers and public health activists. The problem of antibiotic use in farm animals needs to be more widely recognized and carefully evaluated in relation to whatever economic gains may be claimed. As Harold Neu, professor of medicine at Columbia University, says in concluding his 1992 article "The Crisis in Antibiotic Resistance," "The responsibility of reducing resistance lies with the physician who uses antimicrobial agents and with patients who demand antibiotics when the illness is viral and when antibiotics are not indicated. It is also critical for the pharmaceutical industry not to promote inappropriate use of antibiotics for humans or for animals because this selective pressure has been what has brought us to this crisis." Such advice is unlikely to be heeded. As Matt Ridley and Bobbi Low point out in a recent article in The Atlantic Monthly, moral exhortations for the good of the many are often welcomed but rarely acted upon. To get people to cooperate for the good of the whole requires sanctions that make lack of cooperation expensive.

Viruses don't have the same kind of metabolic machinery as bacteria and are not controllable by fungal antibiotics, but there are drugs that can combat them. An important recent example is zidovudine (AZT), used to delay the onset of AIDS in HIV-infected individuals. Unfortunately, AZT, like antibiotics, is not as reliable as it once was because some HIV strains are now (no surprise) resistant to AZT. HIV is a retrovirus, a really minimal sort of organism with special limitations and special strengths. It has no DNA of its own. Its minute RNA code acts by slowly subverting the DNA-replicating machinery of the host to make copies of itself. The cells it exploits include those of the immune system. The virus can hide inside these cells, where it is largely invulnerable to the host's antibodies.

A retrovirus's lack of self-contained proliferation machinery is both its weakness and its strength. It reproduces and evolves more slowly than DNA viruses or bacteria. Another weakness is its low level of reproductive precision, which means that it produces an appreciable number of defective copies of itself. This functional weakness can be an evolutionary strength, however, because some of the defective copies may be better at evading the host's immune system or antiviral drugs. Another strength of retroviruses is their lack of any easily exploited Achilles' heel in their simple makeup.

It takes months or years for HIV to evolve resistance to AZT, in marked contrast to the few weeks it takes bacteria to evolve significant levels of resistance to some antibiotics. Unfortunately, HIV has a long time to evolve in any given host. A single infection, after years of replication, mutation, and selection, can result in a diverse mixture of competing strains of the virus within a single host. The predominant strains will be those best able to compete with whatever difficulties must be overcome (e.g., AZT or other drug). They will be the ones that most rapidly divert host resources to their own use—in other words, the most virulent.

SHORT-TERM EVOLUTION OF VIRULENCE

The evolution of virulence is a widely misunderstood process. Conventional wisdom has it that parasites should always be evolving toward reduced virulence. The reasoning assumes, correctly, that the longer the host lives, the longer the parasites can live and the longer they can disperse offspring to new hosts. Any damage to the host on which they depend will ultimately damage all dependent parasites, and the most successful parasites should be those that help the host in some way. The expected evolutionary sequence starts with a virulent parasite that becomes steadily more benign until finally it may become an important aid to the host's survival.

There are several things wrong with this seemingly reasonable argument. For example, it ignores a pathogen's ultimate requirement of dispersing offspring to new hosts. This dispersal, as noted in the previous chapter, frequently makes use of host defenses, such as coughing and sneezing, that are activated only as a result of appreciable virulence. A rhinovirus that does not stimulate the host to defend itself with abundant secretion of mucus and sneezing is unlikely to reach new hosts.

Another error in the traditional view is the assumption that evolution is a slow process not only on a time scale of generations, but also in absolute time. Such a belief arises from a failure to appreciate the capacity for rapid evolution of any parasite that will go through hundreds or thousands of generations in one host's lifetime. If the virulence of the amoeba that causes dysentery is too low or too high for maximizing its fitness, the virulence can be expected to evolve quickly toward whatever level is currently ideal. We should not expect the present virulence of any pathogen to be in transit from one level to another unless conditions have changed recently. By "recently," we mean last week or last month, not the last ice age, which is what an evolutionary biologist often means by "recently."

Yet another flaw in the conventional wisdom is its neglect of selection among different parasites within hosts, as we just implied in our discussion of HIV. What good would it do a liver fluke to restrain itself so as not to harm the host if that host is about to die of shigellosis? The fluke and the *Shigella* are competing for the same pool of resources within the host, and the one that most ruthlessly exploits that pool will be the winner. Likewise, if there is more than one *Shigella* strain, the one that most effectively converts the host's resources to its own use will disperse the most progeny before the host dies. As a rule, all else being equal, such *within-host selection* favors increased virulence, while *between-host selection* acts to decrease it. A recent comparative study of eleven species of fig wasps and their parasites confirmed that increased opportunities for parasite transmission are associated with increased parasite virulence.

As with many other applications of evolutionary theory, careful quantitative reasoning is needed to understand the balance between natural selection within and between hosts. The graph on the next page is a naive representation of what we have in mind.

An adequate theory of the evolution of virulence must take into account the rate of establishment, in a given host, of new infections; the extent to which these competing pathogens differ in virulence; the rate of origin of new strains by mutation within a host; and the extent to which these new strains differ in virulence. From such considerations it should be possible to infer the expected levels of virulence for a given pathogen, assuming that conditions stay the same, which they never really do. The most important changes would be those that alter the means by which a pathogen reaches new hosts. If dispersal depends not only on a host's survival but also on its mobility, any damage to the host is especially harmful to the pathogen. If you are so sick from a cold that you stay home in bed, you are unlikely to come into contact with many people that your virus might infect. If you feel well enough to be up and about, you may be able to disperse it far and wide. It is very much in a cold virus's interest to avoid making you

FIGURE 4–1. SELECTION WITHIN AND BETWEEN HOSTS.
A shows the effects of an extremely virulent pathogen, which would be
favored by natural selection *within* a host. It exploits its host to maximize
the current rate of dispersal of new individuals to new hosts. It may kill the
host quickly, but while the host lives it does better than any competing
pathogen. *B* shows the effects of a pathogen that is favored by selection
between pathogen communities of different hosts. It maximizes its long-
term total productivity (rate of reproduction times duration, graphically
the area under the production curve). Host death in *B* is most likely from
something other than the pathogen.

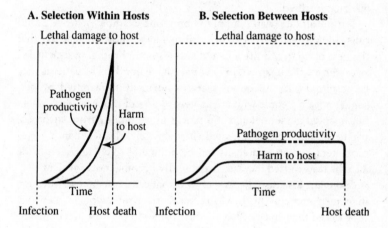

really sick. By contrast, the malaria agent *Plasmodium* gets no benefit
from the host's feeling well. In fact, as shown by experiments with rabbits and mice, a prostrate host is more vulnerable to mosquitoes. People in the throes of a malarial attack are not likely to expend much
effort warding off insects. Mosquitoes can feast on them at leisure and
spread the disease far and wide.

This evolutionary perspective suggests that diseases spread by personal contact should generally be less virulent than those conveyed
by insects or other vectors. Do the facts fit this expectation? They do
indeed. Among Paul Ewald's important discoveries is the truth of
this generalization and its importance for public health. He has
shown that diseases from vector-borne pathogens tend to be more
severe than those spread by personal contact and that mosquito-
borne infections are generally mild in the mosquito and severe in vertebrate hosts. This is to be expected because any harm to the

mosquito would make it less likely to bite another vertebrate. For gastrointestinal pathogens, the death rate is lower for direct, as compared to waterborne, transmission, as long as really sick hosts can effectively contaminate the water supply. As pure water became the norm in the United States early in this century, the deadly *Shigella dysenteriae* was displaced by the less virulent *Shigella flexneri*. As water was purified in South Asia during the middle of the century, the lethal form of cholera was steadily displaced by a more benign form, and the transition took place earliest at the places where water was first purified.

An unsanitary water supply is only one example of what Ewald calls *cultural vectors*. The history of medicine shows repeatedly that the best place to acquire a fatal disease is not a brothel or a crowded sweatshop but a hospital. In hospitals, large numbers of patients may be admitted with infectious diseases normally transmitted by personal contact. People who are acutely ill do not move around much, but hospital personnel and equipment move rapidly from such people to others not yet infected. Inadequately cleaned hands, thermometers, or eating utensils can be quite effective cultural vectors, and the transmitted diseases may rapidly become more virulent.

Take, for instance, the streptococci that can cause uterine infection in women after childbirth. Most nineteenth-century women knew that they risked their lives by having their babies in the hospital, but some still did so. Viennese physician Ignaz Semmelweis noted in 1847 that women in a clinic staffed by medical personnel contracted childbed fever three times as frequently as those in a clinic staffed by midwives. On investigating, he found that doctors came directly from doing autopsies on women who had died from childbed fever to do pelvic examinations on women in labor. Semmelweis proposed that they were transmitting the causative agent and showed that infections were less frequent when examiners washed their hands in a bleach solution. Was he thanked for his wonderful discovery? No. He was dismissed from his post for suggesting that doctors were causing the deaths of patients. He became more and more frantic in his efforts to save the thousands of women who were dying unnecessarily, but he was ignored, and finally, at age forty-seven, he died in an insane asylum. Nowadays, we all accept the need for hygiene in hospitals, but whenever it becomes lax, conditions are perfect for selecting for increased virulence, as in the virulent hospital-acquired (versus community-acquired) infantile diarrhea studied by Paul Ewald.

It is widely believed that HIV is a new pathogen, perhaps originating from a monkey infected with simian immunodeficiency virus (SIV). However, evidence now suggests that monkeys might have acquired SIV from people with HIV. While HIV may have been present in some humans for many generations, AIDS is apparently a new disease, resulting from the evolutionary origin in recent decades of highly virulent HIV strains. AIDS may have arisen because of changed sexual behavior resulting from the socioeconomic disruption of some traditional societies. Large numbers of prostitutes serving hundreds of men per year were so effective at spreading infection that host survival became much less important to virus survival. Those strains that most rapidly exploited their hosts came to prevail within the hosts, and even the highly virulent strains had plenty of opportunity to disperse to new hosts before the old ones died.

In Western countries, AIDS appeared initially as a disease mainly of male homosexuals because their large numbers of sex partners greatly accelerated sexual transmission, and of intravenous drug users because the drug users' needles were effective vectors. As in Africa, the most virulent HIV strains prevailed over the less virulent because between-host selection for lower virulence was greatly weakened. Even highly virulent viruses had abundant opportunities to reach new hosts before the original host died. Conversely, the use of clean needles and condoms can not only curtail the transmission of the virus, *it can also cause the evolution of lower virulence.*

COSTS AND BENEFITS OF
THE IMMUNE RESPONSE

As described in the previous chapter, natural selection has given us a fiendishly effective system of chemical warfare. For every invading pathogen there will be a worst-case scenario as to what kind of molecules it might encounter. Our immune systems have been shaped over a hundred million years to make the pathogen's worst nightmares come true. Unfortunately, every effective weapon can sometimes be dangerous to the one who wields it.

The immune system can make two kinds of mistakes: failing to attack when it should and attacking something when it shouldn't. The first kind of mistake results from inadequate response, so that a disease that should have been nipped in the bud becomes serious. The second kind of mistake results from mounting too aggressive a response to minute chemical differences. Autoimmune diseases such as lupus erythematosus and rheumatoid arthritis could be the result. The average person's degree of sensitivity and responsiveness is presumably close to what has historically been the optimum: enough to counter pathogens but not so great as to attack the body's own structure.

Given that we have this chemical superweapon—immunity—how can we possibly remain vulnerable to infectious diseases? Once again, it is because the infectious agents can evolve rapidly and become better adapted by natural selection. Those variants that are least vulnerable to immunological attack will be those whose genes are best represented in future generations. So the pathogens may evolve one or another kind of defensive superweapon. Molecular mimicry, mentioned in the last chapter, is one such weapon.

ESCALATING DECEPTION

Scientists first developed the concept of mimicry to describe the patterns on butterflies' wings. For instance, the viceroy butterfly looks almost exactly like the monarch butterfly, which birds do not attack because they want to avoid the toxins the monarch caterpillar gets from eating milkweed leaves. The viceroy has no such toxins, but birds mistake it for its bitter lookalike and likewise shun it. Examples are now also known in many other animal groups. Any edible species that by chance resembles a toxic species will have an advantage, and selection will make this *mimic* species look increasingly like the toxic *model*. This is bad for the model because predators that eat the edible mimic learn to go after the model as well. This sets up an arms race between the mimic, which evolves an ever closer resemblance to the model, and the model, which evolves to be as different as possible from its edible neighbors. Some environmental circumstances favor the mimic to such an extent that really detailed resemblances between unrelated species may evolve. We notice such mimicry easily because we perceive so much of the world visually. Detection of chemical mimicry requires more subtle techniques, but there is no reason to think it less common than visual examples.

The molecular mimicry shown by pathogens turns out to be at least as subtle, complicated, and full of surprises as the visual mimicry shown by butterflies and other animals. Deceptive resemblances to human proteins are shown by the surfaces of various parasitic worms, protozoa, and bacteria. If there is any deficiency in the mimicry of human tissues by a bacterium, we can expect it to evolve an improvement rapidly. Pathogen surfaces may have a complex sculpturing of convexities and concavities, and the molecular forms most readily recognized by antibodies are hidden in crevices. As noted in the last chapter, some pathogens alter their exposed molecular structures so rapidly that the host has difficulty producing newly needed antibodies fast enough. This is rapid change without evolution, because the same pathogen genotype codes for a variety of molecular structures.

Mimicry may not only permit pathogens to escape from immunological attack but also make active use of hosts' cellular processes. For instance, streptococcal bacteria make molecules similar to host hormones that have receptor sites on cell membranes. In effect, the bacterium has a key to the lock on the door that normally admits a hormone. Once inside the cell, the bacterium is shielded from immunological and other host defenses. The host has an endosome-lysosome complex that can attack pathogens within its cells, but molecular mimicry and other countermeasures protect the pathogen there too.

NOVEL ENVIRONMENTAL FACTORS

Before leaving infectious disease, we will anticipate a theme of Chapter 10 by noting the large proportion of epidemics that have resulted from novel environmental circumstances. We have already mentioned how changed social conditions may have initiated the AIDS epidemic, but the same is true for many other plagues. Richard Krause, of the National Institutes of Health, reports that early measles and smallpox epidemics spread along caravan routes in the second and third centuries and killed a third of the people in some communities. Bubonic plague, the black death, had long festered in Asia, but became epidemic only when Mongol invaders brought it to unexposed populations in Europe who lived with large populations of flea-infested rats. While we like to imagine that such events are in the past, AIDS continues to spread alarmingly, and the causes of other sudden outbreaks of infection are unknown. The Ebola virus ravaged parts of Africa in the 1980s, killing half of those who became ill, including most of the doctors and nurses who cared for the patients. It stopped as suddenly as it started, for reasons that remain unclear.

Some infectious diseases stem directly from modern technology. Legionnaires' disease arose from an organism that was able to grow and be dispersed from the water in a hotel air-conditioning system. Toxic shock syndrome arose when a new superabsorbent tampon material allowed enough surface area and oxygen for the growth of unusually large concentrations of toxic staphylococcal bacteria. Lyme disease became a problem only when deer populations multiplied adjacent to new suburbs. Influenza has become a major threat since mass worldwide transportation began spreading new strains that contain new genes. It is often called the *Asian flu* because new strains so often originate on Asian farms, where people, ducks, and pigs (some strains are called *swine flu*) live in such close proximity that genes from one influenza strain can easily be passed from one to another.

Tuberculosis became epidemic in Europe with the rise of large, crowded cities. Unsanitary practices and poverty are always cited as causes, but we wonder if the disease didn't become epidemic simply because large numbers of people began spending large amounts of time together indoors. Air exhausted from a TB ward reliably produces infection in guinea pigs but no longer causes infection if it is briefly exposed to ultraviolet light. A single sneeze can produce a million droplets, which settle to the ground at a rate of only about one centimeter per minute in still air. In the open air they would be dispersed or killed by sunlight, but indoors they might last for weeks, as they no doubt did in 1651, when tuberculosis caused 20 percent of all deaths in London.

Finally, we note that epidemics can result from the best of intentions. Polio was not an epidemic disease that caused paralysis until the early twentieth century. Before that time, most children got the disease in the first years of life, when it usually produces only mild effects. By midcentury, improving sanitation delayed the infection until late childhood, when it can be much more severe. Mononucleosis is also less severe at earlier ages. In each of these examples, a disease became a serious problem only when its mode of transmission was changed by novel environments. We will return in Chapter 10 to other novel environmental factors and their role in disease.

INJURY

hen Huck Finn's drunken "Pap" fell over a tub of salt pork and barked both shins, he

fetched the tub a rattling kick. But it warn't good judgment because that was the boot that had a couple of his toes leaking out the front end . . . and the cussing he done then laid over everything he had ever done previous.

Pap acted as if the tub wanted to hurt him, as if kicking and cursing it could deter future harm to his shins. But the kicking and cussing were wasted effort. The tub was not a rival trying to steal Pap's mate, a predator trying to catch him, or even a microorganism stealthily trying to devour his tissues. It was merely inanimate wood.

In discussing such things as tubs of salt pork as sources of injury, we leave behind the conflicting interests, strategies, and arms races that complicate contests between living opponents. The problems associated with injuries are conceptually simpler than those of infectious diseases, but there is complexity aplenty. Some dangers, like being struck by a meteorite, have always been so rare and unpredictable that we have no evolved defenses and can repair the damage only by using general-purpose mechanisms. Others, like exposure to high levels of gamma rays, are so new that we have not had time to evolve adequate defenses. But some dangers, like drowning or attack by predators, have happened often enough in evolutionary history that we have evolved ways to avoid them. This chapter is about the ways we avoid, escape, and repair damage from sources of injury such as mechanical trauma, radiation, burning, and freezing. It is also about why these adaptations do not always work as well as we might wish.

AVOIDANCE OF INJURY

Cooled by milk, the coffee needed to be warmed up just a bit. The microwave oven sounded its three pleasant beeps, and, as one of the authors opened the door, the air filled with the aroma of steaming café au lait. As he grabbed the handle of the ceramic mug, searing pain struck in a fraction of a second, too soon, too intense even to get the hot-handled mug to the counter. It crashed to the floor, splattering hot coffee for yards. After he got his painful hand under cold water, the victim realized that this mug must be different from others, which stay cool to the touch after microwaving. In fact, its handle must have had a metal core. The pain prevented the worse damage that would have resulted from more prolonged contact. The fearful memory of the pain, months later, still makes him shy away from using that particular mug.

Pain and fear are useful, and people who lack them are seriously handicapped. As noted already, the rare individuals who are born without the sense of pain are almost all dead by age thirty. If there are people born without the capacity for fear, you might well look for them in the emergency room or the morgue. We need our pains and our fears. They are normal defenses that warn us of danger. Pain is the signal that tissue is being damaged. It has to be aversive to motivate us to set aside other activities to do whatever is necessary to stop the damage. Fear is a signal that a situation may be dangerous, that some kind of loss or damage is likely, that escape is desirable.

Here we come to a distressing insight. Pain and fear, the sources of so much human suffering, the targets of much medical intervention, are not themselves diseases or impairments but instead are normal components of the body's defenses. Blocking pain and fear in any way other than eliminating the cause may make the damage worse. For instance, people with syringomyelia, a degeneration of the central part of the spinal cord where the pain nerves are located, experience no pain in their hands. A person with syringomyelia would have picked up that hot cup of coffee and drunk it calmly as the flesh curdled on his fingers. If he smokes, his fingers are likely to be charred. Pain is useful, and its link to fear is no accident. When the body is damaged, pain motivates rapid escape and fear prevents recurrence.

But our adaptations for avoiding injury are more subtle than the mere avoidance of pain and its portents. Avoidance can be conditioned more easily to some cues than to others, depending on what kind of harm occurs. Psychologist John Garcia easily conditioned dogs to avoid a peppermint smell associated with gastrointestinal illness but found it much more difficult to use such sickness to condition avoidance of a tone. Dogs also readily learned to avoid an electric shock that was preceded by a tone but had much more difficulty when the cue was an odor. This makes eminent evolutionary sense. Auditory stimuli are more likely than odors to be good cues to the danger of impending injury, while odors are far more reliable indicators of toxic food. Like so many good ideas, Garcia's was difficult to get published, was ridiculed shortly thereafter, and has been praised ever since.

Some cues—for instance, snakes, spiders, and heights—readily elicit fear in ourselves and other primates. It should not surprise us to discover that we instinctively avoid certain cues that have long been associated with such dangers as falling and dangerous animals. After all, a rabbit that learned a fear of foxes only by being bitten would pass on few of its genes. Rabbit brains are preprogrammed to avoid foxes, and it should not be surprising to find that our brains have some similar capacities. But the price of innate behavior is its inflexibility. Better than a fixed innate response would be a more flexible system that induced fear only to stimuli shown to pose a threat. A newborn fawn will stand and stare at an approaching wolf until it sees its mother flee. Then it too flees, and the flight pattern is set for the rest of its life, ready to pass on to the next generation by imitation. Our fears of snakes, spiders, and heights are prepared but not hardwired. They are partly learned and can be unlearned.

Psychologist Susan Mineka carried out an ingenious series of experiments at the University of Wisconsin Primate Center to demonstrate the development of such fears. Monkeys raised in the laboratory have no fear of snakes and will reach over a snake to get a banana. After watching a single video that shows another monkey reacting with alarm to a snake, however, the monkeys develop a lasting phobia of snakes. They will no longer even approach the side of a cage closest to a snake, much less reach across it. By contrast, if the video shows another monkey apparently recoiling in fear from a flower, no phobia to flowers is created, even though the response the monkey sees is otherwise identical. Monkeys readily learn fear of snakes, but not fear of flowers.

GENERALIZED LEARNING AND UNDERSTANDING

In addition to the simple conditioning discussed above, we humans have more subtle adaptations: our capacities for communication, memory, and reasoning. Drivers can imagine that speeding down an icy mountain road is dangerous, even if they have never actually seen it cause an accident. Even those who haven't personally known anyone killed by a fire can understand that a burning building is a serious hazard that a smoke detector can reduce. People can even avoid dangerous things they cannot perceive, such as radon gas, dioxins, and dietary lead, thanks to learning and reasoning. Our capacity to create and manipulate mental representations has many benefits, and the ability to foresee new dangers is clearly one of them. This capacity also helps us to avoid repetitions of actual experiences of danger or injury without creating unnecessary phobias. If we see someone get a shock while wearing suspenders and working carelessly with household wiring, we can reason that the wiring, not the suspenders, caused the misfortune.

REPAIR OF INJURY

Injury cannot always be avoided. Whether at the tenth or the ten thousandth stroke, the hammer eventually comes down on the thumb. The resulting injury brings a whole battery of repair mechanisms into play. Blood platelets secrete clotting factors that soon stem the bleeding, whether external or internal (in the form of a bruise). Other cells secrete a complex variety of substances that cause inflammation, thus raising the temperature of the tissue and making it harder for any invading bacteria to grow. They also keep the thumb painful, thus protecting it from minor stresses that might disrupt the healing process. Simultaneously, the immune system rushes specialized infection fighters to the site. They either attack any bacteria that the injury might have introduced or take them to lymph nodes, where they can be more easily destroyed. Fibrin strands link the tissues together, and, as healing proceeds, they slowly shrink and pull the sides of the wound together. Eventually nerves and blood vessels grow anew into the damaged tissue, and the hammering can proceed as before, albeit more cautiously. These repair processes show a precise, complex coordination that a symphony orchestra might well envy.

Unfortunately, no one has yet written the score for the healing symphony. Many individual parts are described at great length by pathology books, and some attention has even been paid to coordination among the parts, especially the different roles of several groups of immune cells. What we lack is an adaptationist story for the overall process. Such an account would have a plot—the effort to achieve the best possible repairs in as short a time as possible—to which all the details could be related. It would be a tale of optimal trade-offs in the allocation of scarce resources such as time and materials, and between such conflicting values as continued effective use of the damaged part and its protection from stresses that could slow the healing. It would deal with the optimal timing of events, with no job being started until those that must be finished first are completed. It would recognize the need for cooperation and effective communication, not only within such systems such as the immune but also in the participating hormonal, enzymatic, and structural adaptations. It would deal not only with events at the site of injury but with hormonal and other adjustments of emotion and behavior and of physiological processes throughout the body. We hope the score of this well-crafted symphony will be written in the not-too-distant future.

BURNS AND FROSTBITE

Even instantaneous pain was not quick enough to save the tens of thousands of skin cells burned by the hot handle of that coffee mug. Two small regions on the thumb and index finger turned white in seconds. Curdled like an egg white dropped into boiling water, the skin cells formed a mass of denatured protein, a kind of injury more difficult to repair than a minor cut. This is, no doubt, why heat so quickly causes intense pain. Skin with a minor burn heals readily because the mechanism that replaces epidermal cells remains ready to work, but deeper burns pose more difficult problems. If a burn destroys the cells that replace the epidermis, special mechanisms are required to protect the site from infection, clear away the dead tissue, and infuse the region with new skin cells that can grow and gradually resurface the site of the burn. We can do it, but only with time and risk of infection. Far better to avoid the burn.

We have used and abused fire for a hundred thousand years or more. Even before people learned to make fire, they took burning materials from natural sources and maintained fires for cooking and other uses. Has this long association sharpened our reactions to fire's dangers? It would be interesting to learn if we are better defended against hot objects than closely related species are, perhaps by being more sensitive to hot objects or by more rapid healing of burns.

Heat is not the only cause of thermal damage. Freezing can leave cells just as curdled and dead, a condition known as frostbite. Although this was not a routine danger during most of human evolution, it may have shaped our avoidance of extended exposure to cold air and especially to cold water, which is hundreds of times as effective a heat conductor as still air. Liquid nitrogen and dry ice are novel dangers that were entirely absent in the Stone Age. They can be as harmful as fire, but we have not evolved reactions to make us recoil instinctively from liquid nitrogen or dry ice as we do from hot coals.

RADIATION

The most important radiation damage has always been from the sun. Dark-skinned races are fully equipped with the primary defense against the sun's rays, the pigment melanin in the outer skin, which protects the underlying tissues simply by shading them. A few thousand generations of freedom from sunshine, as may happen to animal populations living in caves, results in a loss of the ability to make pigment. The continuous presence of pigmentation in dark-skinned races shows the benefits of its protection against sunshine.

People of European descent pose a special evolutionary problem. Their pale skins show that protection from sunshine has not been such a consistently important factor in their history, and they are especially vulnerable to sunburn. The first warm, sunny days of spring tempt some of them to bare their skins for many hours. Maybe they know from painful experience that this is not wise, but it feels so good after the winter chill. If fear of repeating the previous year's sunburn does not deter them, the pain of this year's will not either, because it comes too late. Only hours after exposure does the sunburned area become sore, red, and feverish. For several days, sheets of dead skin peel off. Recovery can be complete in a week or two, but this may not be the end of the story, because getting even a few serious sunburns greatly increases the risk of skin cancer years or decades later.

Gradually increasing one's exposure to the sun is less harmful, because all but the most fair-skinned individuals can develop a sufficiently protective layer of melanin. Suntan is a fine example of an inducible defense that is developed only when needed. The fact that fair-skinned people are not heavily pigmented all the time suggests that for their ancestors pigment production had important costs to fitness. In Chapter 9 we will explore the possibility that pallor may be adaptive in shady and cloudy environments.

Everyone knows that it is an excess of solar ultraviolet that causes sunburn, but ordinary visible light, while far less destructive, is also photochemically active and potentially harmful. It does not normally harm us, because natural selection has provided almost everyone with enough melanin and enough enzymes that counter photochemical alterations. Organisms that do not ordinarily live with bright illumination are much more sensitive to sunshine or even to some artificial light sources. For instance, when fluorescent lighting first replaced incandescent light in trout hatcheries, it caused massive mortality in trout eggs. Hatchery biologists knew that in nature such eggs develop under a shady layer of streambed gravel. They hypothesized that the mortality resulted from the greater brightness and shorter (blue) wavelengths of fluorescent light. Experiments showed that this explanation was right: when the trout eggs were shielded from the harmful rays, they did just fine.

Sunlight kills skin cells not by thermal damage but by photochemical alteration of essential substances. The resulting abnormal compounds and dead cells invite attack by the immune system. To some extent this is desirable. It is wasteful to devote resources to supporting dead or inevitably dying cells that ought to be efficiently cleared away. It is equally important not to eliminate cells that can adequately repair themselves. Distinguishing between these categories may not be easy. For an injury that doesn't involve pathogen invasion, such as sunburn or perhaps a simple fracture, it may be best to suppress some aspects of the immune response so as not to interfere with healing.

The immune cells themselves, like any others, can be damaged by radiation. At the moment it is not at all clear which of the ultraviolet-induced changes in the immune system are adaptive adjustments and which are impairments. The Langerhans cells in the epidermis, which take up foreign substances and present them to the immune system, react to the ultraviolet wavelengths from 290 to 320 nanometers (UV-B) in complex ways. These cells are intimately associated with nerves that secrete a hormone that blocks their action. UV-B radiation depletes the skin of these cells, thus blocking its ability to react to contact with foreign proteins. Such a lack of sensitivity is characteristic of almost all people who get skin cancer. But UV-B is not the only culprit. There is some evidence that some commercially available sunscreen lotions block UV-B and prevent sunburn but still allow the passage of the longer-wave UV-A, which may damage the skin's immune cells. People who get a rash from being in the sun are often advised to use sunscreens, but sunscreens might in fact make the problem worse by encouraging more exposure to UV-A than they could otherwise tolerate.

An alarming increase in the occurrence of melanoma, a potentially fatal skin cancer, is causing a justified fear of excessive exposure to the sun. The rates in Scotland have doubled in the past decade, and the rates among fair-skinned people are increasing at a rate of 7 percent a year in many countries. Explanations for the increase range from the new cultural desire to be tan to the thinning of the ozone shield, which has always blocked much ultraviolet light. While both of these factors need to be considered, an evolutionary view suggests other explanations too. We do spend more time at beaches, but we spend far less walking in the sun without clothes on. The loss of ultraviolet blocking resulting from ozone depletion is more than counterbalanced in most areas by the local air pollution. What is new is not sun exposure or ozone inadequacy but our *pattern* of sun exposure. People now spend most of their time indoors and then go outside on weekends for intense bouts of unaccustomed exposure. People who are outdoors for hours every day adapt to their amount of usual exposure and are unlikely to get sunburnt. The risk of melanoma is related more closely to the number of sunburns than to the total amount of time spent in the sun.

Another novel environmental factor is the use of chemically complex sun lotions. Blocking ultraviolet radiation does curtail the development of cancerous lesions. A recent controlled study of 588 Australians found that those who used an active sunscreen developed significantly fewer precancerous skin lesions than those who used a cream that did not block much ultraviolet light. But might the chemicals in sunscreens also cause problems? They don't just sit on the surface of the skin but are absorbed into it. What effects do they have on skin cells, and how might they be transformed after binding to tissue proteins and being bombarded by strong light? The answers are very much in doubt. How ironic it would be if we were to discover that skin cancer can be caused, directly or indirectly, by suntan lotions! Attention should also be given to the products used to inhibit the inflammatory process of sunburn. Such inhibition might prevent cancer by preventing unnecessary damage from autoimmune reactions, but it might also protect damaged and potentially cancerous cells from being naturally destroyed by the immune system.

We emphasize that these are not facts but mere speculations that arise from our lack of understanding. Why do we understand so little about sunburn despite the abundance of available information? Understanding that provides a reliable basis for protection and therapy will be reached when researchers well versed in evolutionary reasoning and with a detailed knowledge of the cellular and molecular events of sunburn put together an explanation that: (1) distinguishes UV impairment of skin function from its adaptive responses to UV stress; (2) distinguishes UV impairment of the immune function from the adaptive immune response; (3) distinguishes impairment of Langerhans cell function from adaptive responses; (4) delineates the special components of the repair processes and their coordination; and (5) shows the positive and negative effects of protective lotions applied before exposure and anti-inflammatory medications used afterward.

Sun damage also appears to contribute to cataracts, a clouding of the lens in the eye. While most sunglasses now block ultraviolet light, older models often did not. Instead they reduced the total amount of visible light, so that the pupil actually opened more widely and admitted more ultraviolet light. Worse yet, many of the cheap sunglasses that children are likely to wear still transmit large proportions of the ultraviolet. We wonder whether some of today's cataract patients might owe their misfortune to sunglasses they wore decades ago.

REGENERATION OF BODY PARTS

C hildren often ask the most intelligent questions. "Why," asks an inquisitive child, "can't Uncle Bob grow a new leg like a starfish does?" Why not indeed? If lizards regrow lost tails, starfish lost arms, and fish lost fins, why can we not even regenerate a lost finger? It is remarkable that this question seldom bothers adults, even biologists. The answer, in general evolutionary terms, is that natural selection will not maintain capacities that are unlikely to be useful or that have costs that would exceed the expected benefits. Thus, as noted in Chapter 3, serious damage to the brain or heart was uniformly fatal before the era of modern medicine, and the ability to regenerate these tissues could not be selected for. An individual who lost an arm in a Stone Age accident could bleed to death in a few minutes. If the bleeding were somehow controlled, the victim would likely soon die of tetanus, gangrene, or other infection. Any process that might have allowed our remote ancestors' arms to regenerate has gradually been lost by the accumulation of mutations that have not been selected against.

But what about the loss of a finger? This would not be as likely to cause death as the loss of a whole arm, and such injuries often do heal under Stone Age conditions. Why not regenerate the finger instead of merely healing the wound? The explanation given in the previous paragraph will not suffice here. We suggest instead two other factors. The first is merely that this regenerative ability would not be used very often and would not produce a major benefit. Most people do not lose fingers, and if they do, the long-term impairment need not be serious. A nine-fingered Neanderthal might live to the ripe old age of fifty. Another reason, which we have already repeatedly emphasized, is that every adaptation has costs. The capacity to regenerate damaged tissue demands not only the cost of maintaining the machinery to make this possible but also the cost of a decreased ability to control harmful growths. A mechanism that allows cell replication increases the risk of cancer. It is dangerous to let mature, specialized tissues have more than the minimum needed capability to repair likely injuries, as we will discuss in the chapter on cancer.

A different kind of explanation is often offered for our inability to regenerate a missing finger. Regeneration would require growth hormones, control of cell movement, and many other processes, and they are simply not there. This is another way of saying that, after an early stage of fetal development, the machinery needed for producing a finger is missing. This is the sort of proximate explanation, based on the details of the mechanism, that most medical researchers would think of first. But we also need an evolutionary explanation of *why* the needed machinery is missing, whatever that machinery might be. Such an evolutionary explanation is more likely to satisfy a child's curiosity, and it can lead researchers to fruitful ideas on what sort of repair machinery we might expect to be activated by the loss of a finger. We suggest that the machinery will conform to an optimal trade-off between the advantages of rapid and reliable repair, the costs of the needed machinery, and the dangers of cancer.

6

TOXINS: NEW, OLD, AND EVERYWHERE

"Nat," says Don Birnham (Ray Milland) to his bartender in *The Lost Weekend*, "You don't approve of drinking. Shrinks my liver, doesn't it? It pickles my kidneys. Yes, but what does it do to my mind?" We will consider the effects on his mind in later chapters. Here we will merely mention some effects prior to those on his liver and kidneys.

Don's rye whisky rewards him with a gentle burning sensation as it passes through his esophagus and on to his stomach. His nerves are signaling the deaths of millions of cells as alcohol diffuses rapidly through the usually protective barrier of mucus and enters those cells. If a cell gets more than a critical concentration of alcohol, it dies. Dead cells, or even those with damaged membranes, release *wound hormones* and *growth factors*, which diffuse to other cells held in reserve for just such an emergency. These reserve cells, deep in the protected crypts of the stomach lining, react to the chemical messages by migrating to the site of injury and dividing to produce new cells of the kind needed there. The most exposed layer of stomach cells can be replaced in mere minutes—but does Don allow them enough time before quaffing again?

NATURAL AND UNNATURAL TOXINS

High-proof alcohol is only one of the many novel hazards to which we are exposed. Agricultural pests are controlled mainly by insecticides that did not exist before 1940. Silos are perfused with poisonous vapors to protect grain from insects and rodents. Demonstrably toxic chemicals such as nitrates are used to extend the shelf life of our foods. Many workers inhale toxic dust or fumes, and suburbanites spray insecticides such as lindane into their trees, often with little regard to the possible effects on themselves or their neighbors. There are heavy metals in our water, pollutants in our air, and radon gas rising from our basements. Obviously our modern age is especially hazardous, with respect to poisons in the food we eat and the air we breathe. Right?

Wrong. While we are now exposed to many toxins that did not exist in even the recent past, our exposure to many natural toxins has greatly decreased since the Stone Age and early agricultural times. Recall from the chapters on infectious disease that the contest between consumer and consumed can generate an evolutionary arms race. Plants can't protect themselves by running away, so they use chemical warfare instead. People have always known that some plants are toxic. Gardening books routinely list plants known to have caused illness or death from being eaten. These lists merely deal with the worst offenders. Most plants contain toxins that would be harmful if eaten in more than a minimal amount. Scientists have only recently realized that the toxic substances are not by-products that just happen to be toxic to certain potential consumers; they are the plants' essential defenses against animals that want to eat them (herbivores), and they play a key role in the ecology of natural communities. People who live in the eastern United States needn't look far for an example. Most lawns there are of tall fescue, a grass species popular because it grows fast and resists pests. The fantasy of getting rid of our lawn mowers and letting horses graze our lawns once a week is appealing, but the horses would soon get sick. Most tall fescue is infected at its base with a fungus that makes potent toxins. The grass protects itself by transporting these toxins to the tips of the blades of grass, the perfect location for discouraging herbivores. Tall fescue and its fungus help each other.

Only very recently have a few pioneers, such as Timothy Johns and Bruce Ames and his collaborators, made us aware of the enormous medical importance of the plant-herbivore arms race. We can heartily recommend Johns's book *With Bitter Herbs Thou Shalt Eat It* for an introduction to the role of plant toxins in human history.

Here we are again dealing with an arms race, this time between animals such as ourselves, who eat plants, and the plants, which need to protect themselves from being eaten. When Stone Age inhabitants of central Europe died of starvation late one winter instead of happily filling up on oak buds and acorns, they were losers in the contest with oak trees. Oak buds and acorns are loaded with nutrients, but, unfortunately for potential consumers, they are also loaded with tannins, alkaloids, and other defensive toxins. Early Europeans who filled up on unprocessed oak tissues died even sooner than their starving companions did.

Animals that eat other animals may have to deal with venoms or other harmful materials manufactured by their prey, and they will certainly have to deal with at least traces of the plant toxins eaten by the prey. The monarch butterfly caterpillar, mentioned earlier, feeds on milkweed not only because it has machinery that makes it invulnerable to the milkweed's deadly cardiac glycosides but also because it becomes poisonous itself by consuming the plant and is therefore avoided by potential predators. Many insects and arthropods protect themselves with venoms and poisons. Many amphibians are poisonous, especially the bright-colored frogs that Amazonian peoples use to poison their arrowheads. The vivid colors and patterns of such poisonous animals protect them from predators, who have learned from bitter experience that such prey are not pleasant food items. If you are starving in a rain forest, eat the camouflaged frog that is hiding in the vegetation, not the bright one sitting resplendent on a nearby branch.

How do plant toxins work? They do whatever will keep herbivores from eating the plants. Why are there so many different toxins? Herbivores would quickly find a way around any one defense, so the arms race creates many different ones. The list of different toxins and their diverse actions is impressive. Some plants make precursors of cyanide, which is released either by enzymes in the plant or by the intestinal bacteria of the consumer. The bitter almond is noteworthy in this regard, but apple and apricot seeds use the same strategy, as do cassava roots, which are used for food in many cultures.

All adaptations, however, have costs, and plants' defensive chemicals have theirs. Toxin manufacture requires materials and energy, and the toxins may be dangerous to the plant that produces them. In general, a plant can have high toxin levels or rapid growth, but not both. To put it from the herbivore's point of view, rapidly growing plant tissues are usually better food than stable or slowly growing structures. This is why leaves are more vulnerable than bark and why the first leaves of spring are especially vulnerable to caterpillars and other pests.

Seeds are often especially poisonous, because their destruction would thwart the plant's reproductive strategy. Fruits, however, are bright, aromatic packets of sugars and other nutrients specifically designed to be attractive food for animals that can disperse the seeds contained in them. The seeds within the fruit are designed either to be discarded intact (like peach pits) or to pass safely through an intestinal tract (like raspberry seeds) to be deposited at some distant place surrounded by natural fertilizer. If the fruit is eaten before the seeds are ready, the whole investment is wasted, so many plants make potent poisons to discourage consumption of immature fruits, thus the proverbial stomachache caused by green apples. Nectar is likewise designed to be eaten, but only by whatever pollinators are best for the plant that makes it. Nectar is an elaborate cocktail of sugar and dilute poisons. The recipe has evolved as an optimal trade-off between the need to repel the wrong visitors and not discourage the right ones.

Nuts represent a still different strategy. Their hard shells protect them from many animals, and some, like acorns, are also protected by high levels of tannin and other toxins. Though many acorns are eaten, some are trampled into the ground, while others are buried by squirrels and thus have a chance to sprout into new trees. It takes such elaborate processing to turn acorns into human food that we wonder if the tannin may be too much even for squirrels. Perhaps it leaches out when acorns are buried in moist soil. If so, the squirrels are processing as well as hiding their food, a neat ploy in their arms race with the oak. If you find yourself starving in an unknown wilderness, seek your nourishment in soft sweet fruits, the nuts with the hardest shells, and perhaps some inaccessible tubers. Avoid seemingly unprotected fleshy plant materials like leaves; they are much more likely to be poisonous, as they must be to protect them from your own or any other hungry mouth.

Plants' escalations of the arms race are numerous and varied. Some plants make little defensive toxin until they are mechanically damaged, after which toxin rapidly accumulates in or near the injured part. Damage to a tomato or potato leaf induces production of toxins (proteinase inhibitors) not only at the site of the wound but throughout the plant. A plant has no nervous system, but it does have electrical signaling and a hormone system that can keep all its parts informed about what takes place in a small region. Some aspen trees have even more impressive communication. When a leaf is damaged, a volatile compound (methyl jasmonate) evaporating from the wound can turn on the proteinase response in nearby leaves, even those on other trees. The usual result of such defenses is that insects are discouraged after feeding even briefly. Some particularly adept insects, however, begin their meal by cutting the main supply vein to a leaf so the plant cannot deliver more toxins. And so the arms race goes on.

DEFENSES AGAINST NATURAL TOXINS

The best defenses are, of course, the sorts of avoidance and expulsion already discussed in relation to infectious diseases. We avoid eating moldy bread or rotten meat, which smell and taste bad, because we react with an adaptive disgust to the toxins produced by fungi and bacteria. We rapidly expel toxic substances by spitting or vomiting or diarrhea. We quickly learn to avoid whatever gives us nausea or diarrhea.

Many swallowed toxins can be denatured by stomach acid and digestive enzymes. The stomach lining is covered with a mucous layer that protects it from ingested toxins and stomach acid. If some cells become contaminated, the effect is temporary since stomach and intestinal cells, like those of the skin, are shed regularly. If toxins are absorbed by the stomach or intestine, they are taken by the portal vein directly to the liver, our most important detoxification organ. There, enzymes alter some toxic molecules to render them harmless and bind others to molecules excreted in the bile back into the intestine. Toxin molecules in sufficiently low concentration will be quickly taken up by receptors on cells in the liver and rapidly processed by the liver's detoxification enzymes.

For instance, our protection against cyanide depends on an enzyme called rhodanase, which adds a sulfur atom to cyanide to form a chemical called thiocyanate. Although thiocyanate is far less toxic than cyanide, it still prevents the normal uptake of iodine into thyroid tissue and thus can cause the overworked thyroid gland to enlarge—a condition called goiter. Plants from the genus *Brassica* (including broccoli, Brussels sprouts, cauliflower, and cabbage) get their strong taste from allylisothiocyanate. The ability to taste a related compound, phenylthiocarbamate (PTC) varies greatly, as is well known by generations of students who have tasted a bit of PTC-impregnated filter paper as part of an experiment to demonstrate genetic variation. While some people can't taste PTC, those with a different gene experience it as bitter. They may have an advantage in avoiding natural compounds that cause goiter. About 70 percent of individuals in most populations can taste PTC, but in the Andes, where such compounds are especially likely in the diet, 93 percent of the native people can taste it.

Oxalate is another common plant defense. Found in especially high concentrations in rhubarb leaves, it binds metals, especially calcium. The majority of kidney stones are composed of calcium oxalate, and doctors have for years recommended that such patients keep their diets low in calcium. However, a study of 45,619 men, published in 1992, showed a higher risk of kidney stones for those who had *low* calcium intakes. How is this possible? Dietary calcium binds oxalate in the gut so that it cannot be absorbed. If dietary levels of calcium are too low, some oxalate is left free to enter the body. If, as researchers S. B. Eaton and D. A. Nelson have argued, the amount of calcium in the average diet is now less than half of what it was in the Stone Age, our current susceptibility to kidney stones may result from this abnormal aspect of our modern environment, which makes us especially vulnerable to oxalate.

There are dozens of other classes of toxins, each with its own way of interfering with bodily function. Plants in the foxglove and milkweed family make glycosides (e.g., digitalis), which interfere with the transmission of electrical impulses needed for maintaining normal heart rhythm. Lectins cause blood cells to clump and block capillaries. Many plants make substances that interfere with the nervous system—opioids in poppies, caffeine in coffee beans, cocaine in the coca leaf. Are such medically useful substances really toxins? The dose of caffeine contained in a few coffee beans may give us a pleasant buzz, but imagine the effect of the same dose on a mouse! Potatoes contain diazepam (Valium), but in amounts too small even to cause relaxation in humans. Other plants have toxins that cause cancer or genetic damage, sun sensitivity, liver damage—you name it. The plant-herbivore arms race has created weapons and defenses of enormous power and diversity.

What happens if we overload our bodies with so many toxin molecules that all the processing sites in the liver are occupied? Unlike the orderly queues of shoppers in the supermarket, these molecules do not just wait their turn to be processed. The excess toxins circulate through the body, doing damage wherever they can. While our bodies cannot instantly make additional detoxification enzymes, many toxins stimulate increased enzyme production in preparation for the next challenge. When medications induce these enzymes, this may hasten the destruction of other medications in the body, thus necessitating dose adjustments. Timothy Johns's book notes the interesting possibility that inadequate exposure to everyday toxins may leave our enzyme systems unprepared to handle a normal toxic load when one occurs. Perhaps with toxins, as with sun exposure, our bodies can adapt to chronic threats but not to occasional ones.

Grazers and browsers limit their consumption of certain plants to avoid overloading any one kind of detoxification machinery. This dietary diversification also helps to provide adequate supplies of vitamins and other trace nutrients. Left to our own devices in a natural environment, we do the same. If your favorite vegetable is broccoli and you were given an unlimited supply of it and nothing else, you would not eat as much as you would if given both broccoli and cucumbers. Many weight-loss diets are based on the principle that we eat less if given only a few foods than we would if we had access to a well-stocked cafeteria. We minimize the damage caused by dietary toxins by this instinctive diversification, as well as with our own special array of detoxification enzymes. These enzymes are not as potent or diverse as those of a goat or a deer, but they are more formidable than those of a dog or cat. We would be seriously poisoned if we ate a deer's diet of leaves and acorns, just as a dog or cat would quickly sicken on what we might regard as a wholesome salad.

We can also, more than any other species, protect ourselves from being poisoned by learning about how to avoid it. Only we can read about the dangerous plants in our gardens and woodlands, and we are the species whose diets are most shaped by social learning. A food our mothers fed us can usually be accepted as safe and nourishing. What our friends eat without apparent harm is at least worth a try. What they avoid we would be wise to treat cautiously.

More broadly, there is great wisdom in our innate tendency to follow the seemingly arbitrary dictates of culture. The rituals of many societies require that corn be processed with alkali before it is eaten. Can't you just imagine prehistoric Olmec teenagers ridiculing their elders for going to all the bother? But those teenagers who ate only unprocessed corn would have developed the skin and neurological abnormalities characteristic of pellagra. Neither rebels nor elders could have known that boiling corn with alkali balances the amino acid composition and frees the vitamin niacin, which prevents pellagra, but the cultural practice accomplished what was needed, despite the lack of scientific understanding.

Or consider the prehistoric residents of California, whose main sustenance came from acorns. The abundant tannins in acorns are astringent and combine strongly with proteins, properties that make them especially useful as leather-tanning agents. As noted above, they are highly toxic as they come from the tree. Whether the tannins evolved to protect the acorn against large animals or against insects and fungi is uncertain, but dietary concentrations of over 8 percent are fatal to rats. The tannin concentrations in acorns can reach 9 percent, and this explains why we cannot eat acorns raw. The Pomo Indians of California mixed unprocessed acorn meal with a certain kind of red clay to make bread. The clay bound enough of the tannin to make the bread palatable. Other groups boiled the acorns to extract the tannin. Our enzyme systems can apparently cope with low concentrations of tannin, and many of us like its taste in tea and red wine. Small amounts of tannin may even be helpful by stimulating production of the digestive enzyme trypsin.

Human diets expanded after fire was domesticated. Because heat detoxifies many of the most potent plant poisons, cooking makes it possible for us to eat foods that would otherwise poison us. The cyanogenetic glycosides in arum leaves and roots are destroyed by heat, so that arum could be cooked and eaten by early Europeans. Unfortunately, some toxins are stable at high temperatures, while other new toxins are actually produced by cooking. That tasty char on barbecued chicken contains enough toxic nitrosamines for several authorities to recommend restricting our intake of grilled meat to prevent stomach cancer. Have we been cooking meat long enough to have developed specific defenses against the char toxins? Cooking may have been invented hundreds of thousands of years ago, and it must have started with barbecues on open fires. It would be interesting to know if we are more resistant to heat-produced toxins than our closest primate relatives are.

Since the invention of agriculture we have been selectively breeding plants to overcome their evolved defenses. Berry bushes were bred for reduced spininess and the berries for reduced toxin concentrations. The history of potato domestication, as described in Johns's book, is especially instructive. Most wild species of potato are highly toxic, as you might expect, given that they are an otherwise unprotected, concentrated source of nourishment. Potatoes are from the same plant family as deadly nightshade and contain harmful amounts of the highly toxic chemicals solanidine and tomatidine. Up to 15 percent of their protein is designed to block enzymes that digest proteins. Still, a few wild species can be eaten in limited quantity, and edibility can be increased by freezing, leaching out the toxins, and cooking. We enjoy thoroughly edible potatoes today thanks mainly to many centuries of selective breeding by native farmers in the Andes.

Concerns about pesticides have recently spurred programs to breed plants that are naturally resistant to insects. This protection is provided, of course, by increased levels of natural toxins. A new variety of disease-resistant potato was recently introduced that did not need pesticide protection, but it had to be withdrawn from the market when it was found to make people ill. Sure enough, the symptoms were caused by the same natural toxins the Andean farmers had spent centuries breeding out. An evolutionary view suggests that new breeds of disease-resistant plants should be treated as cautiously as artificial pesticides are.

NOVEL TOXINS

One reason to stress the prevalence of toxins in our natural environment, and our evolutionary adaptation to them, is to provide a perspective on the medical significance of novel toxins. Novel toxins are a special problem not because artificial pesticides such as DDT are intrinsically more harmful than natural ones but because some of them are extremely different chemically from those with which we are adapted to cope. We have no enzymatic machinery designed to deal with PCBs or organic mercury complexes. Our livers are ready and waiting for many plant toxins, but they don't know what to do with some novel substances. Furthermore, we have no natural inclination to avoid some novel toxins. Evolution equipped us with the ability to smell or taste common natural toxins and the motivation to avoid such smells and tastes. In psychological jargon, the natural toxins tend to be aversive stimuli. But we have no such machinery to protect us from many artificial toxins, like DDT, that are odorless and tasteless. The same is true of potentially mutagenic or carcinogenic radioisotopes. Sugar synthesized from radioactive hydrogen or carbon tastes as sweet as that made with ordinary stable isotopes, but we have no way of detecting its dangers.

It is not always easy to tell what the effects of a novel environmental factor may be. For instance, the debate about the possible dangers of mercury in dental fillings has gone back and forth, but Anne Summers and her colleagues at the University of Georgia have recently found that mercury fillings increase the number of gut bacteria that are resistant to common antibiotics, apparently because the mercury acts as a selective factor for bacterial genes that protect against mercury and some of these same genes confer resistance to antibiotics. The clinical significance of this finding is uncertain, but it nicely illustrates the unexpected means by which novel toxins can affect our health.

Since we can no longer, in our modern chemical environment, rely on our natural reactions to tell us which substances are harmful and which are not, we often rely on public agencies to assess the dangers and take measures to protect us from them. It is important to avoid unrealistic expectations of such agencies. Tests on rats are of limited reliability as models for human capabilities, and there are many political difficulties that can frustrate public action on environmental hazards. Scientifically illiterate legislatures can pass laws saying that no amount of any chemical that causes cancer can be allowed in food, even though many such chemicals are already present naturally in many foods. Conversely, political pressures can lead to inadequate controls on known toxins, from nicotine to dioxins. There is no such thing as a diet without toxins. The diets of all our ancestors, like those of today, were compromises between costs and benefits. This is one of the less welcome conclusions that arise from an evolutionary view of medicine.

MUTAGENS AND TERATOGENS

Mutagens are chemicals that cause mutations, which may cause cancer or damage genes and thus lead to health problems for many generations. Teratogens are chemicals that interfere with normal tissue development and cause birth defects. Mutagens and teratogens are not sharply separate from each other or from toxins with short-range effects. Ionizing radiation and mutagens such as formaldehyde and nitrosamines can all cause distress immediately or cancer or birth defects years later.

While it is important to learn which poisons harm everyone, people vary in their susceptibility to many substances, such that one man's meat may be another's poison. We will deal with special aspects of individual variability in the chapter on allergy. Vulnerability varies by age and sex. It seems particularly unlikely that detoxification capability is the same in both adults and the very young, especially during embryonic and fetal development. There are abundant theoretical reasons, as well as data from many experimental studies, that show that actively metabolizing tissues are more vulnerable to toxins than dormant ones, cells that divide rapidly more than quiescent ones, and cells that differentiate into specialized types more than those that merely reproduce more of the same.

All these perspectives suggest that embryonic and fetal tissues may be harmed by lower concentrations of toxins than adult tissues are. We regard Figure 6-1 as a likely picture of vulnerability through human prenatal development. Vulnerability rises rapidly from the level characteristic of a quiescent egg in an ovary to a peak in the critical stages of organ formation and tissue differentiation, then slowly declines to closer to the adult level of tolerance at full term.

We will return to this graph in a moment, but first let's look at a classic mystery of traditional medicine. So-called morning sickness is often the first reliable sign of pregnancy, especially for women who recognize it from prior experience. This nausea and its associated lethargy and food aversions are so common as to be considered a normal part of pregnancy, although they are quite variable in intensity. For some women they mean many weeks of misery, while others aren't bothered much. We may even think of morning sickness as one of the symptoms of pregnancy, as if pregnancy were a disease. The current clinical approach seems to be: pregnancy sickness makes

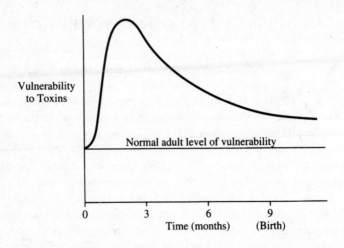

Vulnerability
to Toxins

Normal adult level of vulnerability

0   3   6   9
Time (months)   (Birth)

FIGURE 6–1. Toxin vulnerability at different prenatal ages.

women distressed, so let's find a way to alleviate the symptoms and
make them feel better. Unfortunately, making people feel better does
not always improve their health or secure other long-term interests.
As pointed out in Chapters 1 and 2, natural selection has no mandate
to make people happy, and our long-range interests are often well
served by aversive experiences. Before we block the expression of a
symptom, we should first try to understand its origin and possible
functions.

Fortunately, a biologist thoroughly committed to the adaptationist program has recently wondered at the mystery of morning sickness and devised an explanation. Margie Profet, an independent
scholar and biologist in Seattle, argues that a condition as common
and spontaneous as pregnancy sickness is unlikely to be pathological.
Note on the graph how fetal vulnerability corresponds almost exactly
to the course of pregnancy sickness. This concordance provided
Profet with a crucial clue. Nausea and food aversions during pregnancy evolved, she argues, to impose dietary restrictions on the
mother and thereby minimize fetal exposure to toxins. The fetus is a
minor nutritional burden on the mother in the early weeks of pregnancy, and a healthy, well-nourished woman can often afford to eat
less. The food she is inclined to eat is usually bland and without the
strong odors and flavors provided by toxic compounds. She avoids
not only spicy plant toxins but also those produced by fungal and

88

bacterial decomposition. A lamb chop that smells fine to a man may smell putrid and repulsive to his pregnant wife.

Profet amassed diverse evidence in support of her theory. One example is the correlation between toxin concentrations and the tastes and odors that cause revulsion. Another is the observation that women who have no pregnancy nausea are more likely to miscarry or to bear children with birth defects. Much more evidence needs to be gathered on the evolutionary and related medical questions. We think it unlikely, for instance, that the phenomenon is uniquely human. Is it found in mammals in general, especially herbivores? Do newly pregnant rabbits eat less and choose their food more carefully than either before pregnancy or later on? Studies of wild animals would be the best way to answer these evolutionary questions. The medically more important research can be carried out on laboratory animals. An essential premise to be tested is that some toxins of trivial importance for normal adults have seriously deleterious effects on fetal development. We also need to know the common environmental toxins that are most likely to harm a fetus. We also need to look for associations between diet during pregnancy and the more frequent kinds of birth defects, as well as at individual variations in detoxification enzymes.

Some practical applications of this theory are illustrated by the history of the antinausea medication Bendectin. Pregnant women, understandably, often ask their physicians to do something about their nausea. Recognizing the dangers of drug administration during pregnancy, physicians were generally cautious, but the drug Bendectin was thought to be safe and was widely prescribed. After the thalidomide tragedy, there were many studies on the possible harmful effects of Bendectin, and the equivocal evidence has even been the topic of Supreme Court deliberations. Unfortunately, none of the studies has ever considered the possible functions of morning sickness. Perhaps anything that suppresses morning sickness may cause birth defects indirectly by encouraging harmful dietary choices.

If Profet's theory is correct, it means that pregnant women should be extremely wary of all drugs, both therapeutic and recreational. Fetal alcohol syndrome is perhaps the biggest current problem, affecting thousands of babies every year. Cigarettes can also cause problems, and coffee, spices, and strong-tasting foods may well best be avoided. Certainly, it would be wise to avoid taking any medications if possible. Studies can determine which medications cause major birth defects, but because others may have more subtle effects, it is better to be safe than sorry.

Other than avoiding toxins, what should a pregnant woman do about her nausea? The easy and obvious answer is "Respect it. Your reactions to food are probably adaptive for your baby. Do not succumb to the urgings of others to eat what you are inclined to avoid. Better to offend the host at the party than to risk imposing a long-term impairment on your child." But what about your own suffering? It would be easy enough for two male authors to say, "Accept your nausea; it contributes to your long-term desire for a healthy family." We realize that this is not a satisfactory recommendation. Relief of unpleasant symptoms is desirable as long as side effects are acceptable. We would hope that obstetricians someday will be able to provide their patients with a list of all the substances they ought to avoid. Armed with this knowledge, women could safely use a medication to prevent nausea if it is possible to find one that is effective and to have confidence that it is safe.

People in many cultures, especially pregnant women, eat certain kinds of clay. Although this clay has often been regarded as a mineral supplement, it can relieve gastrointestinal distress and for this reason is used in some modern antidiarrheal medications. Certain kinds of clay, as mentioned in the discussion about acorns, tightly bind soluble organic molecules, including many toxins. In other words, they may relieve symptoms in the best possible way—by removing the harmful cause. Unfortunately, we doubt that it is possible to patent clay. Our present system of drug marketing makes it unlikely that any company would invest the millions needed to test such a product and bring it to market if it could not control an exclusive patent. Regulatory agencies protect us, but they also constrain us.

As fetuses grow older, they become children who tend to hate vegetables. They especially dislike strong-flavored vegetables such as onions and broccoli, the very ones that contain high levels of plant toxins. The developmental course of these dislikes offers a clue to their explanation. Even finicky children often begin to experiment with new foods just as they mature into teenagers and their growth nears completion. The evolutionary explanation for this sensitivity may be the benefits, during the Stone Age, of avoiding the most toxic plants during childhood. Modern-day children and adults would both benefit from eating more of our modern low-toxin vegetables, but there may be a good evolutionary explanation for why children steadfastly resist eating their vegetables.

7

GENES AND DISEASE: DEFECTS, QUIRKS, AND COMPROMISES

The medical school lecture hall was surprisingly full for a Monday at eight A.M. The lecture dealt with nearsightedness. As the room darkened, the overhead spotlights glinted off the eyeglasses worn by nearly half the students. "So that's why so many showed up," murmured the professor.

"The facts are clear," he summarized an hour later. "Myopia is caused by excessive growth of the eye. When it gets too long from lens to retina, the focal point remains above the surface of the retina, so that the image is blurred. Refractive lenses, in the form of glasses or contact lenses, can refocus the image a bit further back so we can see clearly, overcoming nature's inexactitude."

Some hands began to wave. "But what causes the eye to grow too long?" asked one student.

"Genes," he said. "It's as simple as that. Some of us were just unlucky enough to get bad genes. If your identical twin is nearsighted, you will almost certainly be also. If your sibling is nearsighted, the likelihood is high, but not as high. Pulling all the figures together, myopia seems to be a genetic disease with a heritability of over eighty percent."

"But how could such genes survive before glasses were invented?" asked another student. "Without my glasses, I wouldn't last a day on the African plains." The class laughed uneasily.

"Well, the genes might be recent mutations," said the professor. "Or perhaps Stone Age myopic people worked in camp sewing and weaving. In any case, the facts make it clear that myopia is a genetic disorder."

"But how could that be?" the student persisted. "The force of selection against it would be enormous. If such a severe defect can persist, then why aren't our bodies riddled with defects?"

"In fact, our bodies don't work very well," the professor said pointedly. "As you have been learning, we are bundles of genetic flaws. The body is a fragile, jury-rigged device. Our job as physicians is to fix Mother Nature's oversights."

The medical students grumbled a bit more among themselves but did not persist further.

What Genes Do

The instructions for making a human body are contained in molecules of DNA, twisted into our twenty-three pairs of chromosomes. We are still learning the details, wonderful almost beyond belief, of how DNA stores and uses information to build a body. Each DNA molecule is like a ladder, with sides made up of alternating units of phosphate and a sugar called deoxyribose. The information is in the rungs, which are composed of pairs of four molecular components with names abbreviated A, C, G, and T. It is hard to comprehend the amount of information in the genetic code. The DNA in a single cell contains a sequence of twelve billion of the A-C-G-T symbols, the amount of information in a small library. If the DNA in a single human cell were untwisted and the molecules put end to end, it would stretch about two meters. If this were multiplied by the ten trillion cells in the body, it would stretch twenty billion kilometers, about the distance to the planet Pluto!

About 95 percent of human DNA is never translated into proteins. The rest can be divided into somewhere in the neighborhood of one hundred thousand functional subunits called genes. Each gene codes for a single protein. How this DNA chain of As, Cs, Gs, and Ts is translated into a protein is the realm of molecular biology, the fast-growing field that may make more changes in our lives than even the discovery of electricity. There are lonely voices crying for attention to the ethical and political implications of these changes, but the message has not yet gotten through to the general public. Soon it will. Already we have drugs made by DNA cloning. Food plants containing bacterial genes are in production. Pioneering experiments are now relieving previously hopeless diseases by inserting replacement genes into human cells. A less welcome possibility is that an insurance company might, as part of a routine blood test, read samples of DNA and thus learn a client's risks for a variety of diseases. Screening for some genetic disorders in the early stages of pregnancy is already routine, giving a mother of an abnormal fetus the option of terminating the pregnancy.

It is 2010, and Mary, a woman who was in elementary school in 1995, has just found out she is pregnant. "Well, you are pregnant, all right, Mary. Congratulations! The nurse will be here in a minute to explain the normal procedures, but I do need to find out if you want the standard gene screen. I presume so."

"Well, what does it involve?"

"The risks are nonexistent these days, but it is expensive unless you have executive-level health benefits."

"We do have the high-benefits package, but what will the tests tell me?"

"The basic screen identifies forty serious genetic diseases, and then you can get the supplement to look for things like nearsightedness and attention deficit disorder and susceptibility to alcoholism. Most people think it's worth it."

"But what if it shows a problem?"

"Yes . . . well . . . then we will have to talk about what to do. Probably you wouldn't want to terminate just for an increased likelihood of alcoholism or something like that, but it is better to know early. At any rate, it is better to find out now rather than after the problem arises, don't you think?"

"Well, I suppose so, but what am I supposed to do if, say, my baby is going to be nearsighted?"

"Well . . ."

I t will be a few years before the comprehensive testing imagined above is available, but we already know the chromosomal locations of many genes and the code sequences of some. The goal of the controversial Human Genome Project is to unravel the entire code, to find the order of As, Cs, Gs, and Ts that make up the hundred thousand or so genes. When we have the code in hand, we will be able to compare the genes of any individual to those in the standard sequence, thus making it much easier to find abnormal genes.

But is there a "normal" human genetic makeup, as our term *standard sequence* might imply? We are not, of course, all identical. About 7 percent of human genes can differ from individual to individual. For most proteins the variation is low, about 2 percent, while for certain groups of enzymes and blood proteins, 28 percent of genes may have multiple versions. Often, as far as we can tell, different versions of the gene function identically. In other cases, one version (one *allele*) is normal, while the other is defective. In many cases the defective allele is *recessive*, meaning that it has no noticeable effect if paired with the normal allele. If the defective allele is *dominant*, however, even one copy will cause disease.

The problem for an evolutionist is to explain why there is genetic disease at all. Was the professor who gave the myopia lecture right? Are our bodies "bundles of genetic flaws" with legions of disease-causing genes that have not been eliminated by natural selection? Not exactly. There are many genetic defects that are so rare that natural selection has not been able to eliminate them, but they cause relatively little disease compared to more common genes that are, paradoxically, selected for even though they cause disease. We will soon explain how genes that cause disease can be selected for, but first we need to consider how genes work and the rare genetic abnormalities.

All it takes is a single error in the DNA of a sperm or an egg, a C instead of an A, or perhaps a single missing T, to cause a fatal genetic disease. Such errors arise from copying mistakes, from chemical damage, or from ionizing radiation. The wonder is that such errors are not more common. It is estimated that the likelihood of any given gene being altered is one in a million per generation. This means that, on average, about 5 percent of us start life with at least one brand-new mutation found in neither parent. In most cases such mutations have no detectable effects; in others they cause minor effects; in a few they are fatal.

As the individual develops from a single cell to an adult with about ten trillion cells, many more mistakes will creep in. Those that occur after most of the cells in the body have formed are likely to have little effect. Many mutations code for a protein that works about as well as the original or for a protein that is not even expressed in the kind of cell that has the mutation. If the mutation is fatal to the cell, even that will likely be of no consequence since there are usually plenty of other cells available to do the same job. A mutation in a single cell can, however, cause major problems if it knocks out some crucial part of the machinery that regulates cell growth and division. It takes only a single cell multiplying out of control to create a tumor that jeopardizes the whole organism. This hazard is countered by the multiple mechanisms discussed in Chapter 12.

Apart from the difficulties arising from an occasional mutation, how can even an enormously long sequence of only four chemical symbols manage to code for a complete human being? We know quite a bit about how DNA reproduces itself, how it produces RNA, how RNA produces protein molecules, and how these molecules combine to produce microscopic chains or two-dimensional sheets. Beyond that is a vast sea of ignorance in which there are scattered islands of understanding. For instance, we know about some cause-effect relationships and even some details of the machinery of hormonal regulation of tissue development. These isolated points of enlightenment, however, are only the beginnings of a general understanding of animal and plant development.

Even though developmental genetics is still largely mysterious, patterns of genetic transmission are well worked out. At conception, each of us got a copy of each gene at each locus on each chromosome from each parent. A single complete complement of genes (collectively a *genome*) is a random sample of a gene from each locus of the two complete genomes of each parent. So each of us, having two parents, must have two copies of each gene, two complete genomes that together constitute the *genotype*. What we observe in organisms is the *phenotype*, the expression of the genotype as influenced in the course of individual development by many subtle environmental factors. Sexual reproduction is a random shuffling of the genotypes of parents to provide the unique genotype of each offspring. If the shuffling, at a particular locus, gives identical copies of the same gene from both parents, the offspring is *homozygous* at that locus. If it gets a different contribution from each parent, it is *heterozygous*.

A gene will have some average effect over the large number of individuals in which it finds itself over the course of generations, but its effect in any given individual may be quite different from the average. Genes interact with one another and with the environment in determining the features of a phenotype. So a sexually produced individual is unique in many ways and may differ strikingly from either parent. The development of one fertilized egg into two offspring (identical twins) is an asexual reproductive process that produces two individuals with the same genotype.

RARE GENES THAT CAUSE DISEASE

Of the thousands of serious genetic diseases, the vast majority are rare, affecting fewer than one in ten thousand people. Most of these diseases result from recessives, genes that don't cause any trouble except in individuals unlucky enough to get two copies, so there is no normal allele at that locus. This misfortune becomes more likely if you marry a relative, who will have more genes identical to yours than a nonrelative will. This is why marriages between close relatives are more likely to produce abnormal babies.

It is hard for natural selection to eliminate a deleterious recessive gene. If, as is likely, people heterozygous for a rare recessive have no disadvantage, the rate of adverse selection may be so small that natural selection cannot depress the gene frequency further. If a gene is present in one in a thousand individuals and people normally marry nonrelatives, then on average only one in a million will be homozygous. Even if all of these unfortunate people die early in life, the effect of selection is weak. In this situation, new mutations can often create the defective gene as fast as natural selection eliminates it, because as the gene frequency decreases, the prevalence of homozygous individuals decreases even faster. A lethal recessive gene that is created by mutation in one out of a million pregnancies will stabilize in frequency at about one in a thousand individuals. This is indeed a situation in which the power of natural selection is limited.

Dominant genes are another matter. If you have even one copy of a dominant gene that causes a disease, you get the disease and, on average, so will half your children. One of the best known such genes causes Huntington's disease. Most people with this disease have no symptoms until their forties, when their memory fades and their muscles begin to twitch. Some of their nerve cells steadily degenerate until these people cannot walk, remember their own names, or care for themselves. This disease is a particularly vivid example because of its devastating effects and because all known cases can be traced to a small number of European families in the 1600s. One of the men migrated to Nova Scotia. The gene and the disease have been passed on to hundreds of his descendants, including the folk singer Woody Guthrie. In the 1860s a Spanish sailor from Germany, Antonio Justo Doria, settled on the western shores of Lake Maracaibo in Venezuela. His descendants now form the greatest concentration of people with Huntington's disease. Steady detective work and fabulous luck have enabled geneticists to pinpoint the Huntington's gene on the short arm of chromosome 4.

This brings us back to the mystery: Why hasn't this devastating gene been eliminated? The answer is that it usually causes little harm before age forty and thus cannot substantially decrease the number of children born to someone who later develops Huntington's disease. In fact, some studies have suggested that women who later develop Huntington's disease may have more than the average number of children. The reproductive rate of men is somewhat decreased, but net selection against the gene in modern societies must be very slight. Studies estimate that one out of twenty thousand people in the United States have the gene for Huntington's disease.

This disease again illustrates a principle emphasized in Chapter 2: natural selection does not select for health, but only for reproductive success. If a gene does not reduce the average number of surviving offspring, it may remain common even if it also causes a devastating illness. There are genes that cause disease but may possibly increase reproductive success (at least in modern societies)—notably the genes that cause manic-depressive illness. During mania some patients become sexually aggressive, while others accomplish feats that make them successful and thus attractive. If a gene increases the rate of successful reproduction—by whatever mechanism—it will spread.

Table 7-1 offers a classification, based on the beneficiary, of genes that cause disease. While there are many diseases that result from mutation and the limitations of natural selection, they account for relatively little sickness. In most cases the story is more complicated and interesting.

TABLE 7-1 BENEFICIARIES OF GENES THAT CAUSE DISEASE

The individual with the gene:
- Costs and benefits at different stages of the life cycle (Chapter 8); DR3 gene causes diabetes but gives an advantage in utero
- Benefits only in certain environments (e.g., G6PD deficiency is beneficial in areas with malaria; certain HLA haplotypes increase susceptibility to some diseases but protect against others)
- Quirks: Benefits (or at least no costs) in the ancestral environment, costs only in a modern environment (this chapter)

Other individuals:
- Heterozygote advantage to individuals with one copy of a gene, costs to individuals with two copies or none (e.g., the sickle-cell gene)
- The fetus at the expense of the mother (e.g., hPL, see Chapter 13)
- The father at the expense of the mother (or vice versa) (e.g., IGF-II, IGF-II receptor; see Chapter 13)
- Sexually antagonistic selection (e.g., hemochromatosis)

The gene at the expense of the individual:
- Outlaw genes that are perpetuated by meiotic drive (e.g. T-locus in mice)

No one:
- Mutations that occur at a rate equal to the selection rate (equilibrium)
- Some genes are especially vulnerable to mutation because they are very large (e.g., muscular dystrophy). Recessive genes are especially difficult to eliminate because as the frequency of the gene decreases, the force of selection decreases even faster
- Genes present in spite of adverse selection (genetic drift or founder effects)

COMMON GENES THAT CAUSE DISEASE

Sickle-cell anemia is the classic example of a disease caused by a gene that is also useful. The gene that causes sickle-cell disease occurs mostly in people from parts of Africa where malaria has been prevalent. A person who is heterozygous for this gene gets substantial protection from malaria because the gene changes the hemoglobin structure in a way that speeds the removal of infected cells from the circulation. Homozygotes, however, get sickle-cell disease. Their red blood cells twist into a crescent or sickle shape that cannot circulate normally, thus causing bleeding, shortness of breath, and pain in bones, muscles, and the abdomen. People with this disease suffer terribly in childhood, and until recently all of them died before reproducing. An individual homozygous for the normal allele has perfectly good red blood corpuscles but lacks the special resistance to malaria. The sickle-cell gene thus illustrates *heterozygote advantage*. Because of their resistance to malaria, heterozygotes are favored over both kinds of homozygotes: Homozygotes for the sickle-cell allele have low fitness resulting from sickle-cell disease, while homozygotes for the normal allele have low fitness resulting from their vulnerability to malaria. The relative strength of these two selective forces determines the allelic frequencies. Thus, a gene that causes a lethal childhood illness and a gene that makes one susceptible to malaria can both be maintained at high frequencies in the population.

While the sickle-cell allele is the most frequently cited example of a gene that is selected for even though it causes disease, it is unusual for three reasons. First, it is not widely distributed, being originally found almost exclusively in people of tropical African descent. Second, the hemoglobin alteration is a simple sort of adaptation. Most adaptations, such as color vision or the capacity for fever, are complex, closely regulated systems whose assembly requires many genes. By contrast, the sickle-cell allele differs from that for normal hemoglobin only by a single T substituted for a single A. When this genetic code is translated into the protein hemoglobin, the amino acid valine ends up where glutamic acid should be. It is this molecular change that gives the blood cell its abnormal shape and other properties. Third, there is extraordinarily strong selection acting on one gene locus. It may well be that heterozygote advantage is common in human populations, but when selection against homozygotes is weak, the effect is hard to demonstrate.

In areas where malaria is rare, you would expect the sickle-cell allele to decrease in frequency. Indeed, African Americans, many of whom have lived in malaria-free regions for ten generations, show a lower sickle-cell frequency than Africans, lower than any admixture with Caucasian genes would explain. It appears that selection has been decreasing the frequency of the sickle-cell gene in regions where malaria is unimportant, as would be expected from evolutionary theory.

Several other inherited blood abnormalities also protect against malaria, the most dramatic being a deficiency of the enzyme glucose-6-phosphate-dehydrogenase (G6PD). Patients with this abnormality get very sick when exposed to oxidizing medications such as quinine, the original and still effective antimalarial drug. When a malarial parasite uses oxygen in a red blood cell, a lack of G6PD causes the cell to burst, thus interfering with the reproduction of the malarial organism. The ability of some malarial parasites to make their own G6PD illustrates the prevalence of the host-parasite arms race.

One in twenty-five northern Europeans has a copy of the recessive gene that causes cystic fibrosis, and 70 percent of cases are accounted for by a single mutant allele (ΔF508). According to Francis Collins, director of the Human Genome Project, this "suggests that there may have been some heterozygote selection or a very strong founder effect for this particular mutation in the northern European population." Exactly what benefits might maintain the frequency for the gene for cystic fibrosis remain unknown, but decreased death from diarrhea has been suggested.

Tay-Sachs disease kills all homozygote individuals before they reproduce but the gene is present in 3 to 11 percent of Ashkenazic Jews. Maintenance of this high a frequency would require an overall reproductive advantage of 6 percent for heterozygotes compared to homozygotes for the normal gene. Data on infection rates and population distributions suggest that the benefit to heterozygotes may have been protection against tuberculosis, historically a major selective force in Ashkenazic Jews. Fragile-X syndrome is still another common genetic disease, which causes mental retardation in about one out of every two thousand males born. For this syndrome there is direct evidence of increased reproductive success of heterozygous women.

University of California physiologist Jared Diamond recently emphasized another mechanism that can explain the unexpectedly high frequency of some genes that cause disease. He says that as many as eight out of ten conceptions end in early abortion or later miscarriage. The majority are never noticed because they occur before or just after implantation of the embryo. If a gene were to decrease the chances of miscarriage even slightly, it could be selected for even if it also increased the risk of developing a disease. Diamond gives the example of childhood-onset diabetes, which can be caused by a gene called DR3. If one parent is heterozygous and the other is homozygous for the normal allele, 50 percent of the babies would be expected to have the DR3 gene, but the observed rate is 66 percent! It seems that the presence of the DR3 gene in a fetus greatly decreases the miscarriage rate and thus it perpetuates itself, despite causing diabetes.

Phenylketonuria (PKU) may be another example of disease caused by a gene maintained by frustrating the mother's uterine selectivity. When homozygous it causes mental retardation because the body cannot handle normal levels of phenylalanine, an amino acid found in many foods. The retardation can be prevented if the child is given a diet free of this common component. PKU is a fine example of a disease that is completely genetic yet whose effects are completely preventable by environmental manipulation. It is so common (one person in a hundred has the gene) that most states require screening at birth. Why is it so common? Like the diabetes-risk gene, the PKU gene seems to reduce the likelihood of miscarriage and thus to perpetuate itself despite causing disease.

OUTLAW GENES

Oxford biologist Richard Dawkins has viewed the body as the gene's way of making more genes. Genes cooperate to form cells, organs, and individuals only because that is the best way of making more copies of themselves. The body's cells are factories, each with specialized functions, that must cooperate in order for the individual to survive and reproduce. There isn't any way for genes to get into the next generation except by doing their part for the whole organism. Or is there? Given the stakes, one would expect that any gambit that would get a gene into the next generation would be used, even if it decreased the viability of the individual. Does this occur?

Certain genes do compete to get into a sperm or egg, even to the detriment of their carriers. There are several examples, the best known being the T-locus gene in mice. Two copies of the abnormal allele are lethal in males, but males with only one copy transmit it to more than 90 percent of their offspring, instead of the usual 50 percent. This is a fine example of an outlaw gene whose actions benefit itself but harm both the individual and the species. We know about it because it produces a striking effect and because we can do carefully controlled experiments on mice. Might there not be minor human defects that owe their existence to a biased transmission of genes from parent to offspring that balances the decrement of fitness from the defect?

One possibility is polycystic ovaries. This disorder, which accounts for 21 percent of all visits to infertility clinics, is characterized by menstrual irregularity, obesity, and signs of masculinization. A recent study found that 80.5 percent of sisters of women with polycystic ovaries were also affected, a number far too high to be explained by an autosomal dominant or an X-linked gene. Researcher William Hague and his colleagues in Adelaide, Australia, have considered the possibilities that the condition results from transmission of DNA in the cytoplasm of the ovum or from genes that distort the process of meiosis in ways that increase their own chances of getting into an egg, a phenomenon called meiotic drive.

GENETIC QUIRKS:
MYOPIA AND MANY OTHERS

The above diseases result from the specific effects of one gene, but susceptibility to many diseases is determined by the complex effects of many genes. Hardly a week goes by without a newspaper report on the genetics of heart disease, breast cancer, or drug abuse. In most of these polygenic diseases we don't know how many genes are responsible or what chromosomes they are on. We know only that the risk increases if close relatives have the disease. Such associations become especially convincing when people who were adopted as infants show closer resemblances to their biological families than to those in which they grew up, thus reducing the likelihood that the similarity is due to environmental factors.

Susceptibility to coronary artery disease is a good example. The risk of having a heart attack depends considerably on genes. A man whose father had a heart attack before the age of fifty-five has a risk of early death from heart attack five times that of other men. Twins with identical genes have heart-attack rates more similar than those of nonidentical twins, even when all the twin pairs share the same environment. Does this mean that heart attacks are caused by a genetic defect? In some cases, yes. Several abnormalities of cholesterol metabolism have been discovered, one of which is an early candidate for treatment by genetic engineering in which a new gene is inserted into the cells of blood vessel walls. But we also know that heart disease results from eating a high-fat diet. Japanese immigrants to the United States who adopt the high-fat diets of this country have heart attacks more than twice as often as their relatives back home. The rate of premature death from heart disease is high enough that natural selection must be steadily weeding out any genes that contribute to the risk. People often want to know what proportion of heart disease results from genes and what proportion from the environment, but this is not the way the question should be asked. To find out why, let's return to the mystery of myopia.

As the professor said, myopia is a genetic disease. If one identical twin has myopia, the other will almost certainly have it. We have also argued that such a harmful genetic defect would not be expected to persist. Yet about 25 percent of Americans have myopia, often so severe that they would have a hard time in a hunter-gatherer society. How well could they avoid predators, fight in a battle, or recognize a face at fifty paces? Recall poor Piggy, the castaway in *Lord of the Flies*, who without his glasses was trapped "behind the luminous wall of his myopia." Given the disadvantage, it is perhaps no surprise that present-day hunter-gatherer populations have a low incidence of myopia. So why is it so common in modern populations?

When we look carefully at the transition from hunter-gatherer to industrial societies, we see that myopia does not result from a new gene. Native people in the Arctic were seldom nearsighted when they were first contacted by Europeans, but when their children began attending school, 25 percent of them became myopic. It would seem that learning to read and prolonged confinement to classrooms may permanently impair the vision of a substantial proportion of children. Why should this be?

Imagine, for a moment, the difficulty of accurately growing an eye. The cornea and the lens have to focus an image exactly on the retina, even as the eyeball grows steadily during childhood. How exact does the length of the eyeball have to be? The leeway is 1 percent of the length of the eyeball, about the thickness of a fingernail. Is it possible to program the growth of the cornea, the lens, and the eyeball so that the image stays exactly in focus? Unlikely. Yet somehow, even as it grows, the eye keeps images in focus. How?

In a series of experiments, scientists at several laboratories are trying to work out the mechanisms that lead to nearsightedness. First, they noted that an eye with a clouded view grows longer than a normal eye, whether the clouding results from inherited disease, from injury, or from wearing foggy glasses. This is the case for chickens, rabbits, some monkeys, and some other animals, as well as humans. Next, they cut the nerve that carries information from the eye to the brain and found that in some species this stopped the excessive growth of the eye. They began to suspect that whenever a blurred image falls onto the retina, the brain sends back a message, in the form of a growth factor, that induces expansion of the eyeball. The clincher: when only one part of the visual field is blurry, only that part of the eye grows. This kind of asymmetrical growth results in astigmatism.

This mechanism is as necessary as it is elegant. In order to ensure coordinated development of the parts of the eye, the brain processes a signal from the retina, detects blurring, and sends back a signal to increase growth at the particular spot where it is needed. When growth is sufficient, the stimulus stops, and growth does too—except in some people. For 25 percent of us, there is something about reading or other close work that causes the eye to keep growing. Perhaps it is the blurred edges of letters or the plane of focus on a book held close with distant objects all around. It seems possible that printing children's books with especially large, sharply defined letters on oversized pages could prevent some nearsightedness.

Myopia is a classic illustration of a disease whose cause is simultaneously strongly genetic and strongly environmental. To become myopic, a person must have both the myopia genotype and exposure to early reading or other close work. Many other diseases also result from complex gene-environment interactions. For instance, some people eat all the fat they want and never get heart disease, while others eat the same amount of fat and drop dead at age forty. Similarly, some people go through all kinds of losses and never become seriously depressed. For others, the loss of a pet can set off a severe episode of melancholia. Remember also the gene-environment interaction in PKU. For such diseases, it is a mistake to ask what proportion of the cause is genetic and what proportion is environmental. They are both completely genetic and completely environmental.

Can conditions such as myopia and clogged arteries be blamed on defective genes? In our current environment the genes that cause these conditions can certainly create a disadvantage, but in the ancestral human environment many of them might have caused no trouble at all or might even have conferred some real benefits. Perhaps hunter-gatherers with the myopia gene have better vision during childhood. A craving for fatty foods might have been thoroughly adaptive in an environment where such foods were scarce. For this reason we prefer to call such genes not defects, but *quirks*. They have no deleterious effects except in people who are exposed to novel environmental influences. Dyslexia may be another example, difficulty in reading not being a problem for hunter-gatherers.

Susceptibility to drug or alcohol addiction likewise depends on historically abnormal conditions. There are strong genetic influences on susceptibility to alcoholism, but they were a relatively modest problem before the reliable availability of beverages with at least several percent alcohol. Before the rise of agriculture and the vintners' and brewers' development of yeast strains tolerant of high alcohol concentrations, these genes probably were no problem at all. It may prove fruitless to search for a "gene for alcoholism." There may be many such genes on different chromosomes that can make a person susceptible to alcoholism. Many of these genes probably have some positive effects—for instance, a tendency to continue pursuing sources of reward despite difficulties, or a tendency to experience strong reinforcement in response to stimulation of certain brain areas. While it may be tempting to postulate genetic defects in people who abuse drugs, we think it is more likely that the genetic factors that influence drug use will turn out to be a diversity of genetic quirks.

Is there even such a thing as a normal human genome? Certainly no one string of DNA code is ideal, with all deviations to be stigmatized as abnormal. While we humans have much in common, our genes are diverse. There is no one ideal type but only the many varied phenotypes that express the diversity of human genes, all competing in varying environments to get copies of themselves into the next generation.

DON'T LET GENES SCARE YOU

There are widespread but totally unjustified fears and pessimism about genetic influences on human disease and behavior. There is an associated pervasive distrust of scientists who recognize and study these influences. To some extent these anti-gene sentiments reflect a more general antagonism to biological and especially evolutionary explanations among social scientists, the general public, and even some medical professionals. Many people suppose that human behavior and any aspects of human disease that arise from human nature are matters to be dealt with entirely by religion or sociopolitical action, not by seeking biological causes and remedies. When they get cancer or heart disease, however, most people become less concerned about such abstractions.

Is it pointless to try to alter biologically inherited conditions? For some reason, this seems to be a widespread assumption. A recent discussion of myopia contrasted a "use-abuse theory," said to imply that the condition was preventable, with a "genetically determined" theory, said to imply the impossibility of prevention. Fortunately, the subsequent discussion supported the idea expressed in this chapter that myopia is indeed genetically determined and also undoubtedly preventable. In fact, the finding that a medical condition is inherited should generally be considered good news. Genetically programmed development is very much a material process and susceptible to material manipulation. It was the study of the genetic cause of PKU that led to the discovery that its effects could be prevented by a diet free of phenylalanine. Studies of the actions of genes, and of their occasional failure to act, are already preventing and curing many diseases. As Melvin Konner observed in 1983, "The discovery of a genetic determination for a disorder may provide the best hope for an environmental treatment of it." Many others have since made the same point.

Studies of the genetic bases of disease deserve every encouragement, and clinical medicine makes good use of information provided by such studies. When a gene acts against the interests of the patient, the physician should act against the gene. As Oxford biologist Richard Dawkins puts it, we should "rebel against the tyranny of the selfish replicators."

8

AGING AS THE FOUNTAIN OF YOUTH

Let's not have a sniffle,
Let's have a bloody good cry.
And always remember the longer you live,
The sooner you'll bloody well die!
                                    —From an old Irish ballad

The plane sat on the Minneapolis runway in the hot June sun of 1970, the air inside stuffy to the point of apprehension. A white-haired woman, about seventy, turned to the young man in the seat to her left.

"Are you a student?" she asked.

"Well, I just graduated from college. Now I'm about to start med school."

"How wonderful, to have the opportunity to save lives, you must look forward to it."

"Well, uh, yes."

The plane lifted off, fresh air blew from the nozzles above, and a typical airplane conversation ensued—hometowns, common acquaintances, the weather. Then the woman paused, turned to the young man, and spoke plaintively.

"Do you know that there is one disease that we really, really need a cure for, one disease worse than all others, one we all get? Do you know what it is?"

"Uh, no. What?"

"What we really need, what I hope you will look for, is a cure for the worst disease, for old age. It is so terrible, it makes me feel so helpless, and no one has found a cure. Please, please, try to find a cure." Then, she turned away, silent, to gaze out the window.

THE MYSTERY OF AGING

Of the many burdens of consciousness, the fact of death is the heaviest. The possibility of untimely death is frightening, but the inevitability of aging and dying casts the longest shadow on human life. Even apart from religious doctrine, humankind's efforts to overcome aging have been impressively persistent. From Ponce de León searching the wilds of Florida for the fountain of youth to *Life* magazine reporters searching out native Georgians in the former Soviet Union who claim to be 150 years old, human hope lives forever. We, however, do not. By age 80, half of us will die; by age 100, 99 percent; and by about age 115, every one of us will be dead, medical breakthroughs and hopeful news stories notwithstanding.

During the past few hundred years, the *average* length of life (life expectancy) in modern societies has steadily increased, but the *maximum* duration of life (life span) has not. Centuries ago a few people may have lived to 115; today this maximum remains about the same. All the wonders of medicine, all the advances in public health have not demonstrably increased the maximum duration of life. If aging is a disease, it seems to be incurable.

Technically, we are not really talking about aging, the process of growing older from birth onward, but *senescence*, the process of bodily deterioration that occurs at older ages. Senescence is not a single process but is manifested in an increased susceptibility to many diseases and a decreasing ability to repair damage. Death rates in the United States are very low at age 10 to 12, about 0.2 per 1000 children per year. The death rate increases slowly to 1.35 per 1000 at age 30, then increases exponentially, doubling every 8 years. As Figure 8-1 shows, by age 90, the death rate is 169 per 1000. A person age 100 has only a one-in-three chance of living another year. Every year the mortality curve becomes steeper, until eventually we all are gone.

Imagine a world in which all causes of premature death have been eliminated, so that all deaths result from the effects of aging. We would live hearty, healthy lives, until, in a sharp peak of a few years centered at age 85, we would nearly all die. Conversely, imagine a world in which senescence is eliminated, so that death rates do not increase with age but remain throughout life at the level for eighteen-year-olds, that is, about one per thousand per year. Some people would still die at all ages, but half the population would live to age 693, and more than 13 percent would live to age 2000! (See Figure 8-2.) Even if death rates were much higher, say the 10 per 1000 estimates for young adults in India in 1900, eliminating the effects of senescence would still give a substantial advantage, with some people living to age 300. From an evolutionist's point of view, an individual who did not senesce would have, to put it mildly, a substantial reproductive advantage.

This brings us to the mystery. If senescence so devastates our fitness, why hasn't natural selection eliminated it? This possibility seems preposterous only because senescence is such an inescapable part of our experience. Consider, however, the miracle of development: from a single cell with forty-six strands of nucleic acid, a body gradually forms, with each of ten trillion cells in the right place, making tissues and organs that function together for the good of the whole. Certainly it should be easier to maintain this body than to form it!

Furthermore, our bodies have remarkable maintenance capacities. Skin and blood cells are replaced every few weeks. Our teeth get replaced once—but why not six times, like those of elephants? Damaged liver tissue can be rapidly replaced. Most wounds heal quickly. Broken bones grow back together. We can replace missing bits of skin and bone and liver, but some tissues, like heart and brain, do not regenerate. There are revealing differences between species in this regard. In some species of lizards, when the tail is cut off, a new one immediately starts growing. Our bodies do have some capacity to repair damage and replace worn-out parts; it is just that this capacity is limited. The body can't maintain itself indefinitely. Why not?

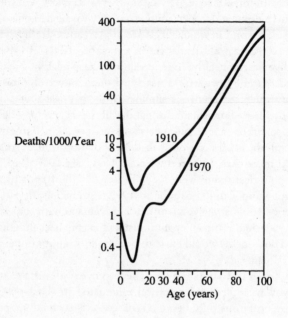

FIGURE 8–1.
The number of deaths per year per 1000 individuals
entering each age is shown at each age for the
United States in the years 1910 and 1970.

What Is Senescence?

For most of us, there is a moment in the mid-forties when we suddenly realize that we can no longer read a book except at arm's length. Yes, some of our hair has fallen out or turned white, and our faces sport some wrinkles, but these changes can be denied far more easily than the weight of a book held on outstretched arms. Fiftieth-birthday parties usually are sickly affairs, where new devotees of mineral water tell nervous jokes about memory loss, hot flashes, and impotence. We know all too well what is to come, but few realize that aging has had a long running start. Senescence starts not at forty or fifty but with far more subtle changes shortly after puberty.

In sports, you don't have to be very old to be past your prime. Look at Figure 8-3, which shows the best times for each age group in running a marathon. The curve looks remarkably like the mortality curves in Figure 8-1. Performance is best in early adult life and thereafter worsens with increasing rapidity. These declines are a sign of senescence. Yes, many people can still run fast at forty, but not as fast as they could at thirty. They would be at a bit of a disadvantage whether chasing an impala or escaping a tiger, and it is the *relative* disadvantage that counts. There is a joke about two men who are running away from a tiger. One stops to put on a pair of running shoes.

"What are you doing that for?" the other asks. "Even with running shoes you can't outrun a tiger."

"No," he says, "but I can outrun you."

The One-Hoss Shay

The "one-hoss shay" in the poem by Oliver Wendell Holmes is the classic metaphor for the remarkable apparent coordination of the effects of senescence. That one-horse carriage . . .

Went to pieces all at once,
All at once and nothing first,
Just as bubbles do when they burst.

111

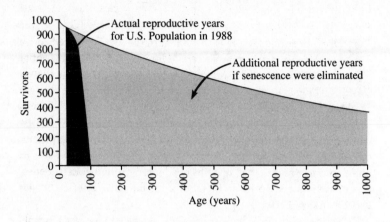

FIGURE 8–2.
Reproductive advantage, if there were no senescence.

Our organ systems also all seem to wear out at about the same rate, on average. Researchers Strehler and Mildvan have measured the reserve capacity of heart, lungs, kidneys, neurons, and other body systems at different ages and found that these diverse bodily systems deteriorate at remarkably similar rates. By the time a person reaches age 100, every system has lost almost all its capacity for meeting increased demands, so that even the tiniest challenge to any system causes a fatal failure. Senescence itself is not a disease but the result of every bodily capacity steadily declining so that we grow steadily more vulnerable to a myriad of diseases, not only cancer and stroke but also infections, autoimmune diseases, and even accidents.

WHY DO WE AGE?

Senescence is a first-class evolutionary mystery. Any explanation must account for the phenomena we've just described. Some clues come from other species. One warm summer evening one of us walked with a group of friends to a picnic on the western shore of Beaver Island in the northern reaches of Lake Michigan. As we mounted the dune overlooking the lake, the last rays of golden sun broke through fiery clouds. We stopped short, breathless at the sight of millions of iridescent wings, flashing in the dying sun. The mayflies formed a golden cloud hovering over the breaking surf, waiting for a chance to mate, lay eggs, and then die on the same day they matured. It seems so wasteful. Yet other species share the mayflies' fate. In the fall, salmon rush up nearby streams, lay their eggs, and die, their rotting bodies washing back to the big lake. This is senescence with a vengeance. How can we understand it?

Many people have thought that senescence must benefit the species. When one of us (Nesse) first became fascinated by senescence as a college sophomore, he investigated every explanation he could find and concluded that senescence was necessary to make room for new individuals so that evolution could keep a species abreast of ecological changes. This was just a step away from the position of the nineteenth-century Darwinian August Weismann, who wrote, in 1881, "Worn-out individuals are not only valueless to the species, but they are even harmful, for they take the place of those which are sound. Hence, by the operation of natural selection, the life of our hypothetically immortal individual, will be shortened by the amount which was useless to the species."

Nagging misgivings about this theory grew after he learned that natural selection acts not for the benefit of the species but normally for the benefit of individuals. There had to be another explanation. When he revealed this preoccupation with the evolutionary explanation of senescence to colleagues in the Evolution and Human Behavior Program at the University of Michigan, they laughed and asked how anyone could possibly not know about the 1957 paper on senescence by a biologist named George Williams.

Williams's paper draws on insights by biologists J. B. S. Haldane and Peter Medawar to show how natural selection can actually select for genes that cause senescence. In 1942, Haldane realized that there would be no selection against genes whose harmful effects occurred only after the oldest age of reproduction. This was a major advance but did not explain why reproduction should cease. In 1946, Medawar went further and showed that the force of selection decreases late in life, when many individuals have been killed by forces other than senescence:

> It is by no means difficult to imagine a genetic endow-
> ment which can favor young animals only at the

expense of their elders; or rather at their own expense when they, themselves, grow old. A gene or combination of genes that promotes this state of affairs will, under certain numerically definable conditions, spread throughout a population simply because the younger animals it favors have, as a group, a relatively large contribution to make to the ancestry of the future population.

Williams expanded these ideas into the pleiotropic theory of senescence. (Genes are called pleiotropic if they have more than one kind of effect.) Imagine that there is a gene that changes calcium metabolism so that bone heals faster, but the same gene also causes slow and steady calcium deposition in the arteries. Such a gene might well be selected for, because many individuals will benefit from its advantages in youth, while few will live long enough to experience the disadvantage of arterial disease in old age. Even if the gene caused everyone to die by age 100, it would still spread if it offered even minor benefits in youth. This argument does not depend on the prior existence of senescence. Other causes of death—accidents, pneumonia, and all the rest—are sufficient to reduce the population at older ages. Nor does the theory depend, like Haldane's, on cessation of reproduction.

The existence of menopause is a related mystery. Why hasn't it been eliminated by natural selection? Menopause is unlikely to be simply a result of senescence because most species continue to have reproductive cycles even into old age and because human menstrual cycles consistently stop within a few years of age fifty instead of gradually tapering off in parallel with other decreases in organ functions. In his 1957 article, Williams offered a possible explanation of menopause. A woman makes a substantial investment in each child, and this investment will pay off genetically only if the child survives to healthy adulthood. If the mother has more babies (with the associated dangers) even as the ravages of age become severe, she is having children she may not be able to care for, and she is risking the future success of her existing children. If, instead, she stops having additional children and devotes her effort to helping those she already has, she may have more total offspring who grow up to reproduce themselves. Recent papers by anthropologists Kim Hill and Alan Rogers challenge this explanation of menopause, but the hypothesis nonetheless offers a fine example of how kin selection might explain apparently useless traits.

Not all genes that cause senescence necessarily have early benefits. Some were simply never exposed to selection because too few people lived long enough in the ancestral environment for the gene to cause a disadvantage. This explanation was thought sufficient by Alex Comfort, the distinguished biologist who is equally well known, in somewhat overlapping circles, for his classic texts *The Biology of Senescence* and *The Joy of Sex*. If Comfort is right, senescence should almost never cause the death of wild animals. He observed that decrepit animals are rarely found in nature and concluded that senescence is not a factor in the mortality of wild populations. But don't forget the sports records. If aging animals run just a little bit slower, they will be caught by predators sooner than their younger competitors are and will thus die from the effects of senescence long before we would notice any decrepitude.

One way to look into this situation is to calculate the force of selection acting on wild populations by comparing the survival curve for the actual population to a curve for an imaginary population that is identical except that its mortality rate does not increase with age. The ratio of the areas under the curves gives an estimate of how much senescence decreases fitness (Figure 8-2 gives an example). In many wild mammals, senescence is a major negative selective force, and most genes that cause senescence are thus within the reach of natural selection. Their prevalence is probably explained by benefits early in life.

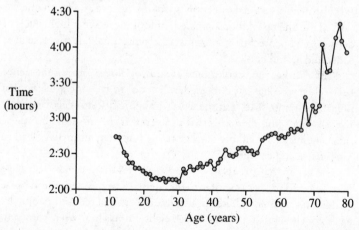

FIGURE 8–3. World record marathon times for men, ages 10 to 79. (Data from *Runner's World*, 1980.)

The astute reader will now want to see some examples of such senescence genes with early benefits. Many genes that have multiple effects are known: for instance, the gene that causes PKU causes fair hair in addition to mental retardation. Here, however, we are interested in genes that have one effect that gives a benefit in youth and another effect that imposes a cost with age. In a 1988 article, University of Michigan physician Roger Albin cited several diseases that may result from such genes. One candidate is hemochromatosis, a disease that causes excess absorption of iron and death in middle age, when the resulting iron deposits destroy the liver. Earlier in life the ability to absorb extra iron may give people with this disorder an advantage (avoiding iron-deficiency anemia) that outweighs the later disadvantage. Albin notes that the prevalence of this gene (about 10 percent of the population has it) can also be explained by heterozygote advantage. Or this may be a gene that is maintained by sexually antagonistic selection. It may benefit women, who need the iron to replace what they lose during menstruation, but harm middle-aged men, who simply accumulate excess iron.

In another example, Albin notes that some people have a gene that results in excess production of a gastric hormone called pepsinogen I. These people are more likely than others to get peptic ulcers and, as they grow older, to die from these ulcers. Throughout life, however, these people have high levels of stomach acid, which may provide extra protection against infection. Insofar as we are aware, no one has carried out the test Albin suggested, of looking to see if high levels of pepsinogen I protect people against gastrointestinal infections such as tuberculosis and cholera.

Paul Turke, an evolutionary anthropologist and senescence researcher who has gone to medical school to become a Darwinian physician, reminds us that the whole immune system is age biased. It releases damaging chemicals that protect us from infection, but these same chemicals inevitably damage tissues and may ultimately lead to senescence and cancer.

The genes that predispose to Alzheimer's disease may also have been selected for because of earlier benefits. The most common cause of devastating mental deterioration, it affects 5 percent of people by age sixty-five and 20 percent by age eighty. It has long been known to be influenced by genetic factors, as shown by many familial cases and by its high frequency in people with three copies of chromosome 21.

In 1993, scientists from the Department of Neurology at Duke University discovered that a gene on chromosome 19 that makes a protein called apolipoprotein E4 is especially common in people who develop Alzheimer's disease. People who are heterozygous for the gene have a 40 percent chance of developing the disease by age eighty. So far as we know, no one has looked for possible benefits early in life in those people who later develop Alzheimer's disease. Now that this gene has been discovered, it should be possible to address the question. S. I. Rapoport at the National Institute on Aging has suggested a related explanation. He notes that Alzheimer's disease is characterized by abnormalities in more recently evolved regions of the brain and that it does not occur in other primates. This led him to suggest that the genetic changes that led to the very rapid increase in human brain size over the past four million years either cause Alzheimer's in some people or produce side effects that have not yet been mediated by other genetic changes. It would be very interesting to see if intelligence early in life is higher, or brain size larger, in people who have the gene that predisposes to Alzheimer's disease.

Considerable laboratory evidence demonstrates that genes with early benefits contribute to senescence. Population biologist Robert Sokal bred flour beetles, those common kitchen pests, and selected for those that reproduced early in their life cycles. After forty generations, the beetles selected for early reproduction produced considerably more offspring sooner in life, but they also aged and died earlier, possibly an effect of genes selected because of their benefits early in the life span despite their costs later in life. Biologists Michael Rose and Brian Charlesworth went the other way, breeding fruit flies that reproduced late in their life cycle. These fruit flies not only had more offspring later in life, they also lived longer and had fewer total offspring, exactly what would be expected if the artificial selection had eliminated genes with early benefits and later costs.

Growing evidence suggests that such genes contribute to senescence in wild animals. For years, gerontologists accepted Alex Comfort's erroneous conclusion that senescence does not occur in wild animals. In a classic example of seeing what they expected to see, many scientists who studied wild populations didn't even bother to check to see if the oldest animals showed increased mortality rates, they just assumed that mortality rates remained constant throughout life. Now that gerontologists have begun looking, however, the evidence is everywhere. For many species, senescence decreases reproductive success more than do all other forces of selection combined. This does not prove the role of pleiotropic genes in senescence, but it certainly challenges the theory that natural selection simply has not had a chance to eliminate the genes that cause senescence.

While evidence for senescence in wild animals supports our trade-off theory of senescence, it has been challenged by evidence that the life span can be readily extended. Severely restricting the diets of rats and mice increases their life span by 30 percent or more. This seems mysterious, because a major increase in life span resulting from something as simple as caloric restriction is inconsistent with our belief that senescence results from many genes acting in concert. So why don't mice and rats eat less and live longer? The first possibility is that they are normally overfed in the laboratory and thus age prematurely. Perhaps their bodies are designed for less lavish diets, so that the starvation experiments were not extending the life span but simply reducing the adverse effects of excess food. This does not seem to be correct. Rats and mice who can eat all they want to are not much heavier than their wild relatives, and poorly nourished rats live even longer than wild animals that are protected from predators and poisons.

Harvard biologist Steven Austad reviewed hundreds of studies of dietary restriction and found the key in a crucial fact mentioned in only a few studies. The food-deprived rats may live longer, but they don't have offspring. In fact, they don't even mate! They seem to remain at a prereproductive state of development, waiting for an adequate food supply. The mechanisms that explain diet-induced longevity remain of great interest, but to an evolutionist, dietary restriction that eliminates reproductive success is no boon but almost as bad as early death.

MECHANISMS OF SENESCENCE

What proximate mechanisms are responsible for senescence and limited longevity? Recent research has found several. Free radicals, for instance, are reactive molecules that damage whatever tissue they contact. Our bodies have developed a number of defenses, especially a compound called *superoxide dismutase* (SOD), that neutralizes free radicals before they can cause much damage. Lack of normal SOD may cause amyotrophic lateral sclerosis (also known as Lou Gehrig's disease), a fatal disease of muscle wasting. The levels of SOD in various species are directly related to their life spans. On the one hand, this shows that damage by free radicals is indeed a proximate cause of senescence, but on the other it demonstrates how natural selection adjusts a defense to whatever level is needed.

Blood levels of uric acid, another antioxidant, are also correlated closely with a species' life span. We humans have lost the ability, possessed by most other mammals, to break down uric acid. Because uric acid crystals precipitate in the joint fluid and cause gout, this loss is often cited in medical books as a deficiency in human biochemistry, but, as noted in this extract from a biochemistry text, it may also be an advantage that facilitates our long life:

> What is the selective advantage of a urate level so high that it teeters on the brink of gout in many people? It turns out that urate has a markedly beneficial action. Urate is a very efficient scavenger of highly reactive and harmful oxygen species—namely hydroxyl radical, superoxide anion, singlet oxygen, and oxygenated heme intermediates in high Fe valence states (+4 and +5). Indeed urate is about as effective as ascorbate as an antioxidant. The increased level of urate in humans compared with prosimians and other lower primates may contribute significantly to the longer life span of humans and to the lower incidence of human cancer.

The flaming painful gouty toe is a cost of a gene that may have been selected because it helps to delay senescence. This gene has effects that are the opposite of those already described, in that the gene gives benefits late in life by slowing aging while exacting its costs throughout adult life. It would be most interesting to see if aging is slower in people with gout.

The levels of an enzyme that repairs abnormal DNA are also higher in longer-lived species. This demonstrates that damage to DNA is a force of selection, and, as with SOD and uric acid, it also demonstrates that nature has found a solution to the problem. If one sees natural selection as a weak force, one sees free radicals and DNA damage as causes of senescence. Appreciation of the strength of natural selection, however, makes one much more inclined to expect that damage from oxygen radicals and defective DNA is limited by evolved mechanisms that are as effective as they need to be to maximize reproductive success.

As Austad points out, the mechanisms of senescence are likely to differ from species to species. Rats and mice, the subjects of most senescence research, are distant from humans, not only phylogenetically but also in their patterns of senescence. Austad therefore proposed extensive cross-species studies of senescence to uncover common patterns. He began his research on an island off the coast of Georgia where opossums had been living without predators for several thousand years and predicted that they would have evolved longer life spans. The fieldwork—catching opossums on both the island and the mainland and determining their ages—took several years. (The task was much easier with the island opossums, because they sleep on the ground in plain view, having lost the defense, essential on the mainland, of hiding all day in deep burrows.) The results of the study? Not only do the island opossums live longer than their landlocked distant cousins, they also age more slowly on a variety of indicators. The cost of these changes, however, is smaller litters at all ages and delayed age at first reproduction. It is clear that the rate of senescence, like other life-history characteristics, is shaped by natural selection.

SEX DIFFERENCES IN RATES OF SENESCENCE

Back to humans. Boys born in the United States in 1985 are expected to live seven years less, on average, than girls, and comparable differences have been found in other countries and in earlier times. Why do women have this advantage over men? The most important evidence for why males age sooner in so many species comes from a cross-species comparison. Males that must compete for mates have shorter lives than females. Part of the increased mortality results from males fighting over females, but even males living alone in cages die sooner than females.

Why are males the vulnerable sex? Male reproductive success is so dependent on competitive ability that male physiology is devoted more to this competition and proportionately less to preservation of the body. Their game of life is played for higher stakes. If unusually fit males can sire large numbers of offspring while mediocre males usually have none, heavy sacrifices must be made in the effort to reach high fitness. Among the processes sacrificed may be those that contribute to longevity.

MEDICAL IMPLICATIONS

Research on senescence seems to be discovering the value of an evolutionary point of view. Gerontologists are realizing that the mechanisms that cause senescence may not be mistakes but compromises carefully wrought by natural selection. An evolutionary view suggests that more than a few genes are involved in senescence and that some of them have functions crucial to life. These genes express their various effects in a seemingly coordinated cluster of escalating signs, because any gene whose deleterious effects occur earlier than those of other genes will be selected against the most strongly. Selection will act on it and other genes to delay its effects until they are in synchrony with those of other genes that cause senescence. This process explains the one-hoss shay effect, the concordance of many signs of senescence even though there is no internal clock that coordinates senescence.

This view discourages the hopes of that lady on the plane, the hope that senescence is a disease that may someday be cured. Hopeful talk about a life-extending research breakthrough is just hopeful talk. What gerontological research does offer, and what justifies considerable investment in studying the mechanisms of senescence, is the likelihood that many diseases of senescence can be postponed or prevented so we can live more fully and vigorously throughout adult life. Despite our pessimism about substantially extending the life span, we concede that the history of science is full of confident theoreticians proving something impossible just a few years before it is accomplished. And we are well aware that natural selection has greatly increased our life span in just a few million years. So we ask not that gerontologists give up their efforts to extend the life span, only that they conduct them in the light of evolution.

We should also note that pessimistic assessments of what science can accomplish often have substantial utility. They provide what philosopher E. T. Whittaker called *postulates of impotence*. Because of such pessimism, engineers no longer try to design perpetual-motion machines and chemists no longer try to turn lead into gold. If gerontologists stop trying to find the fountain of youth in some single, controllable cause of senescence, their efforts may prove more fruitful for human well-being.

The clinician has more immediate concerns. The proportion of people over the age of eighty-five is growing six times faster than the population as a whole. In just the past three decades, the average life expectancy in the United States has gone from 69.7 to 75.2 years. More than a quarter of every health care dollar is now spent on patients in the last year of life, and the need for nursing home beds is expected to quadruple in the next twenty years. Medicine has changed its focus from acute diseases of children and younger adults to chronic diseases of the elderly. Doctors who imagined spending their careers giving antibiotics to stop pneumonia and doing heroic curative surgery now find themselves monitoring high blood pressure, evaluating memory problems, and relieving the symptoms of chronic heart disease. Many of these physicians and their patients still think of senescence as a disease. We expect that knowledge about the evolutionary origins of senescence will have profound effects that are difficult to predict.

This perspective may also change how we see our own lives. Some may find it a consolation to know that senescence is the price we pay for vigor in youth. There is also relief as well as disappointment in knowing that no medical advance is ever likely to extend our lives to any dramatic extent. The search for some pill or exercise or diet that can save us from senescence may be replaced by an appreciation of life as it is, of vigorous function at whatever age. The preoccupation with living forever is likely to be supplanted by a desire to live as fully as possible, while it is possible.

LEGACIES OF EVOLUTIONARY HISTORY

The past! the past! the past!
The past—the dark unfathom'd retrospect!
The teeming gulf—the sleepers and the shadows!
The past—the infinite greatness of the past!
For what is the present after all but a growth out of
the past?
　　　　　—"Passage to India" by Walt Whitman

Phil, the unfortunate television weatherman who lives one day over and over again in the movie *Groundhog Day*, enters a restaurant just as a diner begins to choke on a bite of food. Phil, having observed this scene many times before, calmly steps behind the gasping man, wraps his arms around the man's upper abdomen, and suddenly squeezes hard. The food is expelled from the diner's windpipe and he can breathe again, his life saved by Phil and the Heimlich maneuver.

About one person in a hundred thousand chokes to death each year. While this death rate is small compared to that from automobile accidents, choking has been a persistent cause of death not only throughout human evolution but throughout vertebrate evolution because all vertebrates share the same design flaw: our mouth is below and in front of our nose, but our food-conveying esophagus is behind the air-conveying trachea in our chest, so the tubes must cross in the throat. If food blocks this intersection, air cannot reach our lungs. When we swallow, reflex mechanisms seal off the opening to the trachea so that food does not enter it. Unfortunately, no real-life machinery is perfect. Sometimes the reflex falters and "something goes down the wrong pipe." For this contingency we have a defense, the choking reflex, a precisely coordinated pattern of muscular contractions and tracheal constriction that creates a burst of exhaled air to forcibly expel misdirected food. If this backup mechanism fails and an obstruction blocking the trachea is not dislodged, we die—unless, that is, Phil or someone like him happens to be nearby.

But why do we need the protective mechanisms of traffic control and a backup choking reflex? It would be so much safer and easier if our air and food pathways were completely separate. What functional reason is there for this crisscross? The answer is simple—none at all. The explanation is historical, not functional. Vertebrates from fish to mammals are all saddled with an intersection of the two passages. Other animal groups, such as insects and mollusks, have the more sensible arrangement of complete separation of respiratory and digestive systems.

Our air-food traffic problem got started by a remote ancestor, a minute wormlike animal that fed on microorganisms strained from the water through a sievelike region just behind the mouth. The animal was too small to need a respiratory system. Passive diffusion of dissolved gases between its innermost parts and the surrounding water easily supplied its respiratory needs. Later, as it evolved a larger size, passive diffusion was ever less adequate, and a respiratory system evolved.

If evolution proceeded by implementing sensible plans, the new respiratory system would have been just that, a new system designed from scratch, but evolution does no sensible planning. It always proceeds by just slightly modifying what it already has. The food sieve at the forward end of the digestive system already exposed a large surface area to a flowing current. With no special modifications, it was already serving as a set of gills by providing a large proportion of the needed gaseous exchanges between internal tissues and environment. Additional respiratory capacity was created by slow modifications of this food sieve. Rare minor mutations that made it slightly more effective in respiration were gradually accumulated over evolutionary time. Part of our digestive system was thereby coopted to serve a new function—respiration—and there was no way to anticipate that this would later cause great distress in a Pennsylvania restaurant on Groundhog Day. Today, the food-sieving worm stage in our evolution is still found in the closest invertebrate relatives of modern vertebrates, which have combined respiratory and digestive passages, as shown in Figure 9-1.

Much later, the evolution of air breathing caused some other evolutionary changes that we now have cause to regret. When part of the respiratory region was modified to form a lung, it branched off the lower side of the esophagus that led to the stomach. Accessory openings for air breathing at the surface of the water evolved, understandably, from the already available olfactory organs (nostrils) on the upper surface of the snout, not on the chin or throat. So the air passage opened above the mouth opening and led into the forward part of the digestive tract. Air then passed back through the mouth and larynx to where the trachea branched off and went through this passage to the lungs. This is the lungfish stage (see Figure 9-2).

Subsequent evolution moved the connection from the nostrils back into the throat so that the air passage was as completely separate from the digestive system as it could become without redesigning the structure of the head and throat. Thus a long dual-function passage was gradually shortened until only the crisscross remained, but we and all higher vertebrates are still stuck with it. Vertebrates have the unenviable capacity to be asphyxiated by their food. Darwin pointed out, in 1859, how difficult it is, from a purely functional perspective, to

> understand the strange fact that every particle of food and drink which we swallow has to pass over the orifice of the trachea, with some risk of falling into the lungs, notwithstanding the beautiful contrivance by which the glottis is closed.

We are actually worse off than other mammals because traffic control in our throat is further compromised by modifications to

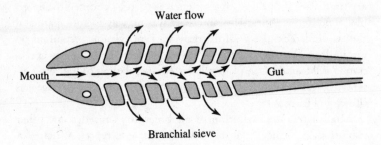

FIGURE 9–1.
Diagram of respiratory and digestive passages of a larval tunicate, and of
the extinct ancestor of all vertebrates, as seen in a horizontal section
through the forward end of the body.

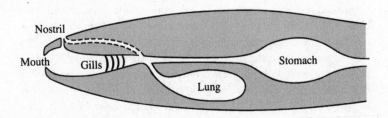

FIGURE 9–2.
The lungfish stage of the evolution of respiratory and digestive systems of
higher vertebrates, as seen in a vertical section to one side of the midline.
The dotted lines show the later shift of the nostril connection to the crossing in the throat, as is found in mammals.

facilitate speech. Did you ever watch a horse drinking? It keeps its mouth in the water and drinks without interrupting its breathing. It can do this because the opening from its nasal region can be precisely lined up with the opening into the trachea. The respiratory passage forms a sort of bridge across the digestive passage, so that when the horse swallows, it can make use of space to the left and right of the bridge. Unfortunately for us, our tracheal opening has slipped further back in the throat, so that the bridge connection can no longer be made. At least not for adults; babies, for the first few months of life, can swallow liquids and breathe simultaneously, like many other mammals. Once they start making the babbling that is the precursor of human speech, however, they can no longer drink like horses. The human capacity for choking represents an ancient maladaptive legacy aggravated by a much later compromise.

OTHER MALFUNCTIONAL
DESIGN FEATURES

Many other serious design flaws make us susceptible to medical problems. Perhaps the most often recognized is the inside-out retina. Vertebrate eyes started as light-sensitive cells under the skin of a minute transparent ancestor. The blood vessels and nerves that served these light-sensitive cells came from the outside, as good a direction as any, for a transparent animal. Now, hundreds of millions of years later, light still must pass through these nerves and blood vessels on the surface of the retina before it reaches the rods and cones that react to the light. The nerve fibers of the retina gather into a bundle, the optic nerve, which must exit the eye to get to the brain. At the hole where the optic nerve exits the retina, there can be no rods and cones. This causes the eye's blind spot. To demonstrate it, close your left eye and focus your right eye straight ahead at the eraser end of a pencil. Move the pencil to the right without letting the eye follow it. The eraser will disappear at a spot about twenty degrees from the forward line of vision. The left eye is similarly blind twenty degrees to the left of its midline.

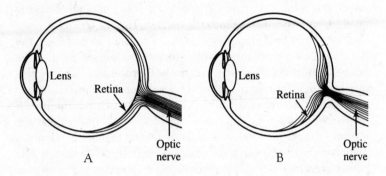

FIGURE 9–3.
A. The human eye as it ought to be, with a squid-like retinal orientation.
B. The human eye as it really is, with nerves and vessels traversing the inside of the retina.

The blood vessels on the retina create another problem. They cast shadows that create a network of blind spots on the retina. To overcome this, our eyes move constantly in tiny twitches so that they scan slightly different areas every fraction of a second. This mass of information is processed in the brain, which compiles it into a coherent image. We are deceived into thinking we see something continuously with both eyes when we may only be seeing it inter-mittently with one. Nevertheless, the shadows, like the blind spot, are always there. To demonstrate this useful self-deception, go into a dark room, press the light end of a penlight against the side of your closed eyelid, turn it on, and gently wiggle it around. When the lineup is exactly right, you will see the shadow of the intricately branching system of parallel veinlets and arterioles that supply the retina.

The inversion of the retina is a universal defect in vertebrates that makes no functional sense. As with the unfortunate intersec-tion between the passages for food and air, the explanation is his-torical, and it applies only to the vertebrates. The functionally analogous eye of a squid has a more sensibly oriented retina with the nerves and blood vessels coming from behind the retina. The squid eye does not need secondary contrivances to minimize the effect of the design flaw that plagues vertebrates, any more than it need worry about eating interfering with breathing. The squid and other mollusks have their own suites of malfunctional historical legacies.

Our inverted retina is responsible not only for slight visual impairment but also for some special medical problems. Any bleeding or minor obstruction of blood flow in the retina casts a shadow that may seriously impair the visual image. Still more serious is the ease with which the light-gathering surface (rods and cones) can lift loose from the underlying interior of the eyeball. Once this condition of *detached retina* gets started, it is a dire emergency that, if untreated, can lead to blindness. The more sensibly designed squid eye, by contrast, has its retina anchored securely from below by numerous nerve fibers so that it cannot become detached.

In addition to those flaws, which affect all vertebrates or all mammals, there are some that affect only humans, or only humans and our closest primate relatives. The appendix is an example. People who recover from appendectomies seem to suffer no disadvantage from not having this part of the human body. The only functional significance of the appendix, as far as we know for sure, is to enable us to have appendicitis. The appendix is the vestige of part of the caecum, a digestive organ in our early mammalian ancestors that helped to process plant foods of low nutritional value. For rabbits and many other mammals, the caecum still serves this function. The shift to a diet of foods with more concentrated nutrition, such as fruit and insects, caused the caecum to degenerate in the course of primate evolution because there was no selection to maintain it. Unfortunately, it has not yet entirely disappeared, and the vestige now makes us vulnerable to appendicitis.

So why does the appendix persist at all? It does make a minor— but by no means important—contribution to the immune system. We also wonder if it might, paradoxically, be maintained by appendicitis. The long, thin shape of the appendix makes it vulnerable when inflammation causes swelling that squeezes the artery to the appendix and cuts off its only blood supply. When filled with bacteria, an appendix without a blood supply cannot defend itself. Bacteria grow rapidly and eventually burst the appendix, spreading infection and toxins throughout the abdominal cavity. A bit of inflammation and swelling is less likely to disrupt the blood supply of a large appendix than that of a long, thin one. Natural selection gradually reduces the size of the useless appendix, but any appendix narrower than a certain diameter becomes more vulnerable to appendicitis. Thus, deaths from appendicitis may paradoxically select for a slightly larger appendix, maintaining this less-than-useless trait. Selection is also almost certainly very slowly making the appendix shorter, but in the meantime the appendix may be maintained by the shortsightedness of natural selection. We wonder if other vestigial traits might also be maintained because further diminishing them increases vulnerability to a disease.

Many primates and most other mammals can make their own vitamin C, but we humans cannot. Our ancestral shift to a high-fruit diet, rich in vitamin C, had the incidental consequence about forty million years ago of allowing the degeneration of the biochemical machinery for making this vitamin. Our frugivorous close relatives share our requirement for dietary vitamin C. All animals need particular organic substances (vitamins) in their food, but different groups have different requirements.

Some of our vulnerability to mechanical damage can also be blamed on various past evolutionary developments. A sharp blow to the side of the human head may fracture the skull, damage the brain, and cause death or permanent impairment. The same blow to an ape head may result merely in a bruised temporalis muscle and temporary impairment of chewing. The difference arises from the increased size of the human brain case and shrinkage of the jaw musculature, which incidentally rob the skull of its earlier cushioning. The hard hats construction workers and cyclists wear are a technological fix for a biological deficiency. If workers and cyclists go on being careless about wearing their hard hats, perhaps in another million years we will again have a thick padding of tissue under our scalps to reduce brain injuries.

The same increased skull size has resulted in a fetal head that fits through a human pelvis only with difficulty. A woman's pelvic structure is slightly different from a man's, so as to provide a large birth passage and, as childbirth approaches, the pubic joint loosens to further facilitate the passage of the infant. Yet childbirth is still more difficult than it would be if the vagina could open outside the massive ring of pelvic bone, perhaps above the pubis on the lower abdomen. The passage of the vagina through the pelvis is a severe historical constraint on the evolution of any further increase in fetal head size. This constraint, of having to fit an oversize head through the pelvic ring of bone, explains why human babies have to be born at such an early and vulnerable stage of development, compared to, for example, ape babies.

The prevalence of maladaptive human design features has been recognized for a long time. A 1941 book by George Estabrooks, *Man, The Mechanical Misfit*, describes many of the structural defects and compromises in human anatomy, especially those that result from turning a horizontal four-footed animal into an upright two-footed one. The weight of the top part of the body greatly compresses the vertebrae in the lower spine, and standing upright requires more muscular effort than a horizontal posture would. The pelvis was originally designed to resist a back-to-belly force of gravity, not the fore-to-aft force that ours must resist as long as we remain upright, either standing or sitting. Elaine Morgan's recent book *The Scars of Evolution* gives a readable account of these maladaptive legacies.

A long list of medical problems, ranging from minor annoyances to serious disabilities, results from the mechanical inadequacies of our adaptations for an upright posture and two-footed locomotion. Perhaps the most important is the episodic lower back pain experienced by so many people. Our knees, ankles, and feet are also extraordinarily vulnerable. How often do we hear of athletes missing games because of knee and ankle injuries? One of the authors once leaped high in a volleyball game, and when he came down only his left foot was on the court. The right landed on the foot of a teammate and turned sharply inward, seriously straining the vulnerable lateral ligament, which is usually the part that fails when an ankle is sprained. The author met his classes on crutches for the next week and was glad he was not part of a band roving over the Paleolithic savannah. He also regretted that the human ankle is not better designed.

The abdominal viscera of a mammal are enclosed in sheets of tissue designed to hang from the upper wall of the abdominal cavity. This is fine for a mammal on all four legs, but in an upright mammal the sheets of tissue may be said to hang from a vertical pole, a grossly ineffective arrangement that causes such diverse problems as digestive system blockages, visceral adhesions, hemorrhoids, and inguinal hernia. The mammalian circulatory system is also compromised by upright posture. It works fine for a dog or a sheep, but our upright posture increases the hydrostatic pressure in the lower extremities and can cause varicose veins and swollen ankles. The opposite effect, deficient blood pressure in the brain, can result in dizziness or momentary partial blackout when we suddenly stand up from a recumbent position.

Sometimes the body's responses to problems are just the opposite of what would be adaptive. When the heart muscle is too weak to pump all the blood it receives, the blood backs up into the lungs and feet and causes shortness of breath, swollen ankles, and other symptoms of congestive heart failure. You might expect that this would cause excretion of excess fluid, but patients with heart failure retain salt and fluid, and this excess blood volume makes the problem even worse. This response is maladaptive in patients with heart disease, but, as internal medicine physician Jennifer Weil points out, the body's response is designed for a different problem. In a natural environment, most instances of deficient blood pumping would result from bleeding or dehydration, in which the fluid retention mechanism would be useful indeed! Heart failure occurs mainly in old age and mechanisms to conserve body fluid can be useful throughout life, so this system is a fine example of a cause of senescence which is maintained because of its benefits in youth.

We have been discussing defects in the basic plan of the human body. These should not be confused with mere inadequacies of execution and random departures from optimal values. As a general rule for any readily measured physical feature, it pays to be in the middle of the pack, as we illustrated previously with the birds with longer- or shorter-than-average wings, which were especially likely to be killed in a storm. Unusually tall or short people tend not to live as long or as healthily as those of average height. Babies of average birth weight are usually better off than those who are much heavier or lighter. Everyone knows that high or low blood pressure is not as good as normal blood pressure. A high level of adaptive performance usually requires that many quantitative characteristics closely approach optimal values. While no individual is perfect, the various parameters sometimes combine to yield remarkable excellence. Yet even in near perfection there is substantial variation—as is well known to those basketball stars who played against Michael Jordan.

Many design features, while not maladaptive, are functionally arbitrary and explicable only as historical legacies. In mammals, the right side of the heart circulates the blood to the lungs, the left side to the rest of the body. In birds it is the other way around, for no better reason than that birds and mammals came from different reptilian ancestors that took arbitrarily different routes to cardiac specialization. Either way works equally well. Some arbitrary features can be advantageously exploited. Many people who are alive today would be dead except for the happenstance of everyone having two kidneys. When one fails or is donated, the other is able to do double service. By the same logic, many people die of having only one heart. The reason we have two kidneys and one heart is simply that, right from their origins, all vertebrates had two kidneys and one heart. This is pure historical legacy and has nothing to do with the advantage of having two of one organ or the disadvantage of having only one of another.

We have belabored what is wrong or arbitrary with the human body because the design flaws can cause many medical problems, but we hope that our readers will also realize that much about it is just right. Our oversize brains may be vulnerable to injury and may impede childbirth, but they make us the unchallenged leaders of the animal kingdom in cognitive capability and in all the social and technological advances that this makes possible. No other species in the history of our planet has ever controlled its environment to the extent that we have since the invention of agriculture. Similarly, our longevity is impressive in relation to that of any other mammal, except a few, such as elephants, that are far larger than we are. We can live about half again as long as any other primate.

Moreover, many of our other adaptations are equal or superior to those in other mammals. Our immune system is superb. Also, despite conspicuous design flaws and individual imperfections, our eyes and related brain structures incorporate layer upon layer of information-processing marvels that extract the maximum amount of usefulness from visual stimuli. If hawks, for example, have visual acuity that is in some ways superior to ours, this one kind of superiority must be purchased with some kind of trade-off. Animals that can see better than we can in the dark cannot see as well in the light. Normal human vision approaches a theoretical maximum of sensitivity and discrimination over a wide range of conditions. We are only beginning to understand how it is that a face, seen from one angle at a certain distance, may later, from another angle and distance, be instantly recognized. No current computer can approach such feats. Our hearing is so sensitive to some frequencies that if it were more sensitive we would not hear as well. Informative sounds would be lost in the noise of random air molecules colliding with our eardrums.

THE FINISHING TOUCH

We have been discussing mainly attributes that humans share with other vertebrates, other mammals, or other primates. Our discussions of our problems with upright stature also apply to extinct members of our genus, *Homo*. We now turn to more explicitly human legacies, with an emphasis on the evolutionary adjustments made in the period from about one hundred thousand to about ten thousand years ago. While natural selection has been changing us in many small ways in the last ten thousand years, this is but a moment on the scale of evolutionary time. Our ancestors of ten thousand or perhaps even fifty thousand years ago looked and acted fully human. If we could magically transport babies from that time and rear them in modern families, we could expect them to grow up into perfectly modern lawyers or farmers or athletes or cocaine addicts.

The point of the rest of this chapter, and the following one, is that we are specifically adapted to Stone Age conditions. These conditions ended a few thousand years ago, but evolution has not had time since then to adapt us to a world of dense populations, modern socioeconomic conditions, low levels of physical activity, and the many other novel aspects of modern environments. We are not referring merely to the world of offices, classrooms, and fast-food restaurants. Life on any primitive farm or in any third-world village may also be thoroughly abnormal for people whose bodies were designed for the world of the Stone Age hunter-gatherer.

Even more specifically, we seem to be adapted to the ecological and socioeconomic conditions experienced by tribal societies living in the semiarid habitat characteristic of sub-Saharan Africa. This is most likely where our species originated and lived for tens of thousands of years and where we spent perhaps 90 percent of our history after becoming fully human and recognizable as the species we are today. Prior to that was a far longer period of evolution in Africa in which our ancestors' skeletal features lead scientists to give them other names, such as *Homo erectus* and *Homo habilis*. Yet even these more remote ancestors walked erect and used their hands for making and using tools. We can only guess at many aspects of their biology. Speech capabilities and social organizations are not apparent in stone artifacts and fossil remains, but there is no reason to doubt that their ways of life were rather similar to those of more recent hunter-gatherers.

Technological advances later allowed our ancestors to invade other habitats and regions, such as deserts, jungles, and forests. Beginning about one hundred thousand years ago, our ancestors began to disperse from Africa to parts of Eurasia, including seasonally frigid regions made habitable by advances in clothing, habitation, and food acquisition and storage. Yet despite the geographic and climatic diversity, people still lived in small tribal groups with hunter-gatherer economies. Grainfield agriculture, with its revolutionary alteration of human diet and socioeconomic systems, was practiced first in southwest Asia about eight thousand years ago and shortly thereafter in Egypt, India, and China. It took another thousand years or more to spread to central and western Europe and tropical Africa and to begin independently in Latin America. Most of our ancestors of a few thousand years ago still lived in bands of hunter-gatherers. We are, in the words of some distinguished American anthropologists, "Stone Agers in the fast lane."

DEATH IN THE STONE AGE

I magine what it must have been like in that idyllic era. You were born into a nomadic band of forty to a hundred people. Whatever its size, it was a stable social group. You grew up in the care of various close relatives. Even if your local band consisted of a hundred or more people, many of them were distant cousins. You knew them all and knew their genetic and marital connections to yourself. Some you loved deeply and they loved you in return. If there were those you did not love, at least you knew what to expect from them, and you knew what everyone expected of you. If you occasionally saw strangers, it was probably at a trading site, and you knew what to expect of them too. In a sparsely peopled world the necessities of life—plant and animal foods uncontaminated by pesticides—were there for the taking. You breathed the pure air and drank the pure water of a preindustrial Eden.

Having asked you to imagine an idyllic past, we now urge that you be more realistic. Like other Golden Age legends, such as the age of chivalry or that delightful antebellum world into which Scarlett O'Hara was born, it is a fabricated myth. Enjoy it in fantasy or fiction, but do not let it mislead serious thought on medicine or human evolution. The unpleasant fact is that our hunter-gatherer ancestors lived with enormous difficulty and hardship. Simple arithmetic on the rates of death and reproduction makes this conclusion inescapable. Death always balanced reproduction, even though people reproduced at something approaching the maximum feasible rate.

In most primitive social systems, women start bearing children as soon as they are able to do so, which, because of nutritional limitations, is often delayed until about age nineteen. Pregnancy and childbirth are followed by two or three years of lactation, which inhibits ovulation. Then the mother is soon pregnant again, whether this is medically advisable or not. In the unlikely event that she remains fully fertile and survives to menopause, she will probably produce about five babies. Having more children would require shortened lactation periods, and this is unlikely given the limited foods available for babies in preagricultural societies.

But even if hunter-gatherer women averaged only four children before succumbing to sterility or death, only half their babies could have survived to maturity. Otherwise the human population would have steadily increased, and this obviously did not happen. Even an increase of 1 percent per century would cause a population to become a thousand times as numerous in less than seventy thousand years, but populations remained extremely sparse until the invention of agriculture. The conclusion is thus quantitatively inescapable that deaths almost precisely kept up with births for nearly all of human history. The extraordinarily low death rates of the last few centuries, and especially in the last few decades in Western societies, show that we live in times of unprecedented safety and prosperity. It is no doubt difficult for most readers of this book to appreciate the harshness and insecurity of human life under natural conditions.

Mortality rates in the Stone Age, like those of today, were highest in infancy and declined throughout childhood. Many early deaths in some groups were from infanticide, motivated by parents' economic hardship or imposed by patriarchs. While fictional accounts of Stone Age conditions probably exaggerate the ravages of predation and other wild-animal attack, lions, hyenas, and venomous snakes were ever-present hazards and took a steady toll, with children especially vulnerable. Death rates from poisoning and accidents were far higher than they are now.

The infectious diseases, which were probably the most important source of mortality for all age groups, were not the same bacterial and viral diseases that afflict us today. Most of today's infections depend on rates of personal contact only possible in abnormally dense populations. Back then, vector-borne protozoa and worms were common causes of prolonged sickness and ultimate death. Many of these diseases are not merely lethal but most unpleasantly so. Some readers will know how unpleasant malaria can be, from personal experience or from knowing someone who has had the disease. It is a lark compared to other protozoan diseases such as kala-azar, which slowly destroys the liver and other viscera; parasites such as lungworms, which cause death by suffocation; hookworms, which are seldom fatal but can make children grow into physically and mentally defective adults; and filaria, which among other things cause elephantiasis. The name comes from the swelling of the limbs and scrotum to elephantine proportions because the parasites block the lymphatic vessels.

Food was often abundant for hunter-gatherers, but memories of bounteous fruit harvests or an occasional big kill must have been a poor solace during the regularly recurring famines. Climatic variations induce fluctuation in resources. Even in the most stable climates, food abundance varies because of plant and animal diseases. Prior to the invention of reliable preservation techniques, temporarily abundant food could not be saved for leaner times. Even foods preserved by drying or smoking could be attacked by pests that could frustrate the most careful planning for future emergencies.

Shortages of vital necessities were not only directly stressful, they also encouraged strife. Imagine that people from a hill tribe were suffering from a protein shortage, while people in the valley were feasting on the abundant fish from their lake. The people from the hills would no doubt insist that their leaders take them to that lake, no matter how loudly the valley people asserted their exclusive fishing rights. If catching the fish means killing the fishermen and appropriating their fishing gear, that is what the hill people might decide to do. Even in the absence of economic necessity, human nature often finds excuses for armed robbery and attendant taking of life. Fortunately for early tribal societies, they lacked the technology of transport and communication that permitted banditry on the scale practiced by Genghis Khan or Alexander of Macedon.

Human nature has, of course, its nobler aspects. There are such things as love and charity and honesty. Unfortunately, the evolutionary origins of such qualities are rooted in their utility in parochial tribal settings. Natural selection clearly favors being kind to close relatives because of their shared genes. It also favors being known to keep one's promises and not cheating members of one's local group or habitual trading partners in other groups. There was, however, never any individual advantage from altruism beyond these local associations. Global human rights is a new idea never favored by evolution during the Stone Age. When Plato urged that one ought to be considerate of all Greeks, not merely all Athenians, it was a controversial idea. Today, humanistic sentiments still face formidable opposition from parochialism and bigotry. In fact, these destructive tendencies are aggravated by what we just now called the "nobler" aspects of human nature. As Michigan biologist Richard Alexander so neatly put it, today's central ethical problem is "within-group amity serving between-group enmity."

LIFE IN THE STONE AGE

Human nature was formed in what anthropologists have recently termed (following a 1966 suggestion by psychiatrist John Bowlby) the *environment of evolutionary adaptedness*, or *EEA*. Despite their frequent reference to the EEA, anthropologists differ widely about what it was like. They cannot directly observe the ways of our ancestors of tens of thousands of years ago or the effects of environmental conditions on the human genetic makeup. They must base their conclusions on indirect evidence: skeletal remains, stone tools, cave paintings, and information about modern groups with seemingly primitive economies and social conditions.

The shortage of information is serious. What are the historically normal conditions of human childbirth? This is just one of many basic questions for which there is no assured answer. We suspect that the correct answer to many such questions is, *it was highly variable.* Attitudes toward childbirth differ enormously among different cultures today, and there is no reason to believe they were any less variable a hundred thousand years ago. They must also have been quite variable within social groups. The solicitude offered to a chief's wife no doubt differed from that proffered a concubine captured from a hostile tribe. Giving birth during times of plenty in a settled camp might have been rather different from giving birth in leaner times or during travel to a new location.

We also suggest that the correct answer to other important questions is, *it varied.* What sorts of rewards went to gifted poets, artists, or others of high intellectual attainment, compared to those who were good hunters or warriors? How stratified, by family connections or merit, were the socioeconomic conditions? Was inheritance matrilineal or patrilineal? What were the child-rearing customs? What were the religious doctrines and constraints, and how strong a factor was religion? These questions would have vastly different answers in different societies in the EEA. There is no one "natural" way of human life.

Despite great variation in the human adaptations to a variety of EEA conditions, the available evidence does support some generalizations. Social systems were constrained by economics and demography. No elaborately stratified societies with hereditary class structures were possible in the Stone Age, because groups that had to gather their food from within walking distance necessarily remained small. Likewise, no chief of a nomadic band can have dozens of wives when the band only includes a few dozen people. Prior to the development of agriculture, no chief could control enough land, wealth, and people to build cathedrals or pyramids.

Social systems were also constrained by the physiological and structural differences between the sexes. The physiological costs of reproduction involved in pregnancy and lactation are borne entirely by women. By what rules were the economic costs of reproduction apportioned? Again, we suggest, *they varied*. On the basis of what we know about current human groups, husbands no doubt contributed significantly in most cultures, but in others a mother's brothers and other relatives made a greater contribution. Likewise, the gross physical differences between the sexes imply behavioral differences. The greater size and strength of men suggest that these attributes provided important competitive advantages, especially in the competition for mates. We discuss this and related matters in Chapter 13.

Economic necessity often demanded that adults and older children of both sexes spend much of their time searching for and preparing food. It is usually assumed that men did the hunting and women the gathering in hunter-gatherer societies, although the antiquity and importance of big-game hunting have been exaggerated in fictional accounts of Stone Age life. Archery and other weapons effective against such animals as deer were in fact not invented until late in the Stone Age. Dogs, which can play crucial roles in many hunting techniques, were not common human associates before about fifteen thousand years ago. Meat or hides from large animals may often have been procured not by hunting but by scavenging or stealing from other predators.

The mainstay foods in the Stone Age would seem to us inedible or too demanding of time and effort. We would find most of the game strong-tasting and extremely tough. Most of us have little appreciation of the tedious skinning and butchering it takes to turn a wild animal carcass into a serving of meat. Many wild fruits, even when fully ripe, are sour to our tastes, and other plant products are bitter or have strong odors. We find them unpleasant thanks to our adaptations that make us avoid toxins, as discussed in Chapter 6. Most natural human foods require a far greater labor of preparation and chewing than the foods we eat now. Domesticated animals and plants have been artificially selected to be tender, nontoxic, and easily processed.

Despite the abundance of foods available in the EEA much of the time, the village elders would have been able to remember times of severe famine. Actual starvation may have been rare, but deaths from the combined stresses of disease, malnutrition, and poisoning by the excessive consumption of marginally edible plants were probably common. These same stresses also would have caused abortion of fetuses, curtailment of lactation, reduced fertility, and actions such as infanticide and the abandonment of the old or impaired.

In addition to xenophobic conflict with other groups, social strife within groups, famines, and toxic diets, there were many other environmental stresses. Our ability to tolerate the atmospheric pollution of modern cities may owe much to our many thousands of years of exposure to smoke toxins from woods and other fuels. Imagine living in a hut with a fire on the floor and only a small hole in the roof. Atmospheric pollution was different in the EEA, but it was substantial and real. We would find the odors of a Stone Age settlement most unpleasant. There were no soaps or deodorants, no flush toilets, or readily cleanable chamber pots, or any installations worthy of the term latrine. Wastes of various kinds were taken away to some customary distance and no further. Other wastes simply accumulated where they were produced. The average Stone Ager lived in a dump and moved away when conditions got really bad.

Children grew up, and adults lived out their lives, in the constant awareness, and sooner or later the personal experience, of woeful illness, painful injury, physical handicaps, debilitation, and death. There were no antibiotics, tetanus shots, or anesthetics, no plaster casts, corrective lenses, or prosthetic devices, no sterile surgery or false teeth. Our remote ancestors had few cavities, but they had many other dental problems. Teeth could be injured or lost in accidents, and they could literally wear out before what we call middle age. Abrasive plant products can wear molars down to gum level, as seen in some fossil skulls and even in some contemporary groups.

Lest it seem that our account of the EEA is merely a selection of items for a catalog of horrors, we should emphasize that we are discussing our fully human ancestors, with a fully human capacity for pleasure as well as pain and a fully human intellect. The bonds of kinship and friendship could be strong and a source of great pleasure and security. In seasons of plenty there would be abundant time for play: games, music and dancing, storytelling and poetry recitals, intellectual and theological disputes, and the creation of ornamental artwork. The cave paintings at Lascaux, France, created perhaps 25,000 years ago, have been described by anthropologist Melvin Konnor as "a Paleolithic Sistine Chapel" that impresses a sensitive observer "whether religious or not—whether expert or not—with a strong sense of the holy." And our ancestors also had the ability to look on the bright side in times of adversity and to find reasons for laughter. Mark Twain's hero Sir Boss in *A Connecticut Yankee in King Arthur's Court* lamented having to listen, at a sixth-century campfire, to the same jokes he had already found tiresome in the nineteenth. We suspect that if he had gone back to the Stone Age he would have groaned at many of the same jokes.

10

DISEASES
OF CIVILIZATION

You have now spent several hours reading this book. Do you realize how much thoroughly abnormal use of your eyes this feat required? Was the light source the sun, with its normal spectrum? Probably not, at least not entirely. How much muscular exertion did you expend during those hours of reading? How could you be so inactive for that much time without jeopardizing your well-being, perhaps your life, by having spent inadequate time and effort in vigilance against enemies and in foraging for food? But you are in fact well fed? How long did it take to pick or dig or hunt or fish for your last meal? How much shelling and grinding and butchering? If the food was cooked, how long did it take you to gather the fuel and kindle the fire? How much sweating and shivering have you done in the last twenty-four hours? What's that about thermostatically controlled heating and air conditioning? How bizarre! And what are the long-term consequences of such meager challenge to your body's built-in temperature controls?

As the last chapter (we hope) made clear, only the grossly uninformed or irrationally romantic would think we were ever better off than we are now. Rousseau's noble savage and the Flintstones' merry capers are delightful in escapist fiction, but the reality was painful and sad compared to our lives today or even to when farming first replaced nomadic scrounging. Agriculture led to urban civilization, with its durable architecture and associated fine arts, and the nautical and other technological advances that permitted exploration of distant lands. The domestication of hoofed animals enabled one worker to do jobs that would previously have required several. It also contributed to revolutionary advances in transportation. Continuing technological advances have led to ever greater freedom from want and freedom of movement for ever larger numbers of people.

The long-term consequences of the soft and gratifying lives we now enjoy are mostly beneficial or harmless, but many of the advantages we enjoy today are mixed blessings. Benefits have costs, and even the most worthwhile benefits can be costly to our health. For a good example we need look no further than the effects of lower mortality rates in early life. Because fewer people die young from smallpox, appendicitis, childbirth complications, and hunting accidents, the death rates from old-age afflictions like cancer and heart failure are much higher now than they were a couple of generations ago. This is largely because a higher proportion of people live to the ages at which the body becomes especially vulnerable to these illnesses. The price of not being eaten by a lion at age ten or thirty may be a heart attack at eighty. Modern practices of food production, medicine, public health, and industrial and household safety have drastically improved the prospects of surviving to old age. Unfortunately, the increased effects of aging are not the only bad aspects of the good life.

Novel environments often interact with previously invisible genetic quirks to cause more variation in phenotypes, some of it outside the normal range. As described already in the chapter on genetics, these abnormalities arise only when a vulnerable genotype encounters an environmental novelty. Novel physical, chemical, biological, and social influences will cause problems for some people and not others or will have different effects on different individuals depending on their specific genetic makeup. We have already discussed some human examples; for instance, the genetic quirks that cause myopia impose problems in literate societies, but they caused no difficulties for our ancestors.

Our ways of getting food changed the environment in ways that created new problems. Thousands of years ago some of our ancestors hunted wild goats or cattle. Hunters followed herds for hours in the hope of killing one of the animals for food, hide, and other resources. Sometimes they may have found, early in the morning, the same herd they had been following the day before. If animals can be followed for two days, why not three, or a week, or a month? How long would this go on before the hunters would start thinking of the herd as their own, driving off wolves or rival groups of hunters or other predators and chasing strays back into the group to maintain a large herd? This process gradually converted hunters into nomadic herdsmen.

Other ancestors were more vegetarian and found that some plants could produce a lot more food if they were intentionally planted for later harvest. Plowing, weeding, fertilizing, and selecting variants with the highest yields soon became standard practice and resulted in steadily greater and more reliable food production. It has been supposed that local increases in population may have encouraged the invention of agriculture or its adoption from neighboring peoples. Whether this is true or not, agriculture permitted the maintenance of much denser and more sedentary populations than could be supported by hunter-gatherer economies. Increased population density then became a source of other problems, some of which will be discussed in this, others in the next four chapters.

MODERN DIETARY INADEQUACIES

Paradoxically, the increased food production made possible by herding and agriculture resulted in nutritional shortages. There are more calories and protein in a bushel of wheat than in a handful of wild berries, but there is more vitamin C in the berries. If wheat provides most of the calories and protein for a farming community, deficiencies of vitamins and other trace nutrients are much more likely to arise than they would be with the more diversified diets of hunter-gatherers. If the wheat or other agricultural produce is also used as feed for the domestic animals that provide meat or eggs or milk, the farmers' meals are much improved, but shortages, especially of vitamin C, remain a threat.

Iceland is a good example, with a vitamin C problem that lasted well into this century. Icelandic farmers raised mainly sheep, which grazed the wild grasses of the countryside. The more successful families might have had a dairy cow, but mutton provided a large part of the diet, and wool was the chief commercial export, sold mostly to Danish colonials. The money so earned allowed the farmers to import flour and such luxuries as coffee and sugar. Nothing in the list so far contains vitamin C, which was provided mainly by blueberries and other wild plant foods. Unfortunately, the supply of these commodities was strictly seasonal. During winter and spring, when diets were notably lacking in vitamin C, many a seemingly robust and healthy Icelandic farmer would start bleeding from the gums and feeling lethargic and depressed, the usual symptoms of scurvy. Some members of a family would sicken and others not, with the severity of scurvy varying greatly.

For those who survived the winter sick with scurvy, folk wisdom came to the rescue. As soon as the marshes thawed, people could dig angelica roots, which are a fair source of vitamin C. The so-called "scurvy grass" might be sprouting at the same time and could be eaten as an alternative. The observation that such wild produce could cure scurvy antedated the use of citrus fruits for preventing the disease among long-distance sailors. Scurvy is a disease of civilization. Before people relied heavily on domestic plants and animals, they never had such abnormal diets as those of Icelandic farmers in the winter or sailors at sea for months at a time.

Long before there were any ocean voyages such as those of the original limeys or those that took the first settlers to Iceland, people suffered from other dietary deficiencies resulting from agriculture. About fifteen hundred years ago, some native tribes of the south central United States abandoned their hunter-gatherer lifestyles and started growing corn and beans. The change is clearly recorded in their skeletal remains. Compared with earlier skeletons, those of the farmers are on average less robust, and they often show effects of nutritional deficiencies of the B vitamins and perhaps protein. Despite these deficiencies, such farmers may have been less likely to die of starvation than their ancestors. They may even have been more fertile, because cornmeal and beans can facilitate earlier weaning. Nonetheless, in important respects, they were not as healthy.

These diseases of civilization thus existed fifteen hundred years ago in what would become Tennessee and Alabama, and long before that in earlier agricultural regions of other continents. The same sorts of nutritional deficiencies afflict the impoverished people of many third-world countries today. Our Stone Age ancestors no doubt faced frequent shortages of food, but if they were getting enough calories they were probably getting enough vitamins and other trace nutrients. Shortages of specific vitamins and minerals arose in just the past ten thousand years or so.

We are now aware of the need for vitamins and minerals, and we get more of them from a modern diversified diet than many early agriculturalists did. Contrary to pharmaceutical sales pitches, few modern people need vitamin supplements. If we eat a diverse array of fruits and vegetables, some of them preferably uncooked, and especially if we also get abundant protein from grains, legumes, and animal products, we are getting all the vitamins, minerals, and other nutrients we need. The current danger for most of us is not the deprivation suffered by our ancestors but an excess of nutrition.

MODERN NUTRITIONAL EXCESSES

A wise man once observed that it makes little sense to worry about excessive eating in the festive week from Christmas to New Year's Day. It makes much more sense to worry about what we eat between New Year's Day and Christmas. Of course, it is possible to overeat in a week. We can even overeat at one sitting, but this was also a danger in the Stone Age, and we are equipped with instincts to avoid doing so. There comes a point at which we feel stuffed and no longer hungry, even for that honey-cured Christmas ham. This normally puts an end to the meal and keeps us, as it did our ancestors, from overburdening the machinery of digestion, detoxification, and assimilation. Modern overnourishment is mainly the result of steady long-term overeating.

In the Stone Age it was adaptive to pick the sweetest fruit available. What happens when you take people with this adaptation and put them in a world full of marshmallows and chocolate eclairs? Many will choose these modern delicacies over an equally available peach, itself sweeter than any fruit available in the Stone Age. Marshmallows and chocolate eclairs exemplify the *supernormal stimuli* described by students of animal behavior. The classic example came from observations on geese. If an egg rolls out of a nest, a brooding goose will reach out and roll it back with her chin. Her adaptive programming is "If a conspicuously egglike object is nearby, I must roll it into the nest." What happens if you put both an egg and a tennis ball near her nest? She prefers the tennis ball. To her it looks more egglike than an egg. There can be supernormal stimuli in any sensory mode, for instance, taste. Next time you find yourself reaching for a slice of apple pie instead of an apple, think of that goose who seems to think she should incubate a tennis ball.

Our dietary problems arise from a mismatch between the tastes evolved for Stone Age conditions and their likely effects today. Fat, sugar, and salt were in short supply through nearly all of our evolutionary history. Almost everyone, most of the time, would have been better off with more of these substances, and it was consistently adaptive to want more and to try to get it. Today most of us can afford to eat more fat, sugar, and salt than is biologically adaptive, more than would ever have been available to our ancestors of a few thousand years ago. Figure 10-1 shows a plausible relationship between intake and benefit of these substances and proposes a contrast in the foraging capabilities of a Stone Age tribesman and of a high-salaried diner in a gourmet restaurant.

An overwhelming amount of preventable disease in modern societies results from the devastating effects of a high-fat diet. Strokes and heart attacks, the greatest causes of early death in some social groups, result from arteries clogged with atherosclerotic lesions.

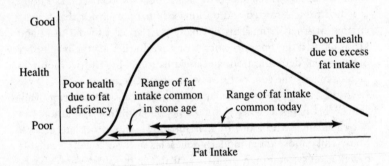

FIGURE 10–1.
Our view of the dependence of health and fitness on resource availability, such as dietary fat intake per month. We propose that fat availability in the Stone Age would seldom exceed the levels indicated. Today an originally adaptive craving for fatty foods may lead to intakes far out on the negative slope to the right.

Cancer rates are increased substantially by high-fat diets. Much diabetes results from the obesity caused by excess fat consumption. Forty percent of the calories in the average American diet come from fat, while the figure for the average hunter-gatherer is less than 20 percent. Some of our ancestors ate lots of meat, but the fat content of wild game is only about 15 percent. The single thing most people can do to most improve their health is to cut the fat content of their diets.

One of us once met with three others early one morning to travel to a hearing on claims that agricultural uses of pesticides were endangering the health of nearby suburban residents. A stop at a diner for breakfast yielded a vivid memory. One of the eaters lamented the likelihood that the wheat and eggs in his pancakes were no doubt contaminated with unnatural pesticides and antibiotics that might give him cancer ten or twenty years later. Perhaps so, but these toxins were a minor danger to his future health compared to the grossly unnatural fat content of his sausage and buttery pancakes, and the enormous caloric value of the syrup in which everything was bathed. The cumulative effect of that kind of eating is surely more likely to cause future health problems than are the traces of exotic chemicals.

Some people are more prone to this sort of overdosing than others. This is indicated by observable variation across the spectrum from underweight to overweight. Overweight people are more likely to suffer the cardiovascular problems associated with excess nutrition and to have higher rates of various cancers. This common impression is supported by recent studies. University of Michigan geneticist James Neel and his associates have noted that efforts to relieve the chronic malnutrition of the Pima Indians of Arizona inadvertently caused an epidemic of obesity and diabetes. He proposed that the affected individuals had what he called "thrifty genotypes," a genetically based ability to get and store food energy with unusual efficiency. With what seem like normal diets many Pimas steadily increase their stores of body fat. This could well be adaptive in a world that threatens frequent famine. Those who have built up copious fat stores might survive a prolonged food shortage while their less efficient associates perish. Thrifty genotypes are not adaptive in a world in which food shortages never occur. The most famine-adapted individuals may just get fatter and fatter until medical problems or other difficulties intervene.

Excess nourishment is not an easily corrected health hazard, and many common solutions may do more harm than good. Voluntary restrictions on food intake may be interpreted by the body's regulatory machinery as a food shortage. The result may be a resetting of the basal metabolism so that calories are used even more efficiently and further fat reserves are amassed. Another consequence of food restriction is intensified hunger, with consequent eating binges. Studies of artificial sweeteners fail to show that they help people to lose weight, a finding that might have been expected. Sweetness in the mouth, throughout human evolution, has reliably predicted sugar in the stomach and shortly thereafter in the bloodstream. It is not surprising that the sweet taste quickly resets metabolic processes so as to curtail the conversion of fat and carbohydrate reserves into blood sugar. This would be adaptive only if, in fact, the stomach contents quickly compensate for the change. If the sugar signal is a lie, there could soon be deficient blood sugar and increased hunger, especially for quick-energy sources like candy. There has been little recognition of such effects of artificial sweeteners. A similar hazard may be anticipated for nonnutritive fat substitutes. There are now desserts that look and taste like ice cream but are not only low in sugar but free of fat. What kind of signals do these send to the metabolic regulatory mechanisms?

Dental cavities are rare in preagricultural societies. If dental workers had been conscious of Stone Age fitness requirements, they would have realized long ago that the twentieth-century epidemic of dental caries must have been due to some environmental novelty, which we now know to be the frequent and prolonged exposure of the teeth to sugar. It nourishes bacteria on the teeth that generate acid, which in turn erodes the dental enamel. Here likewise there is prehistoric evidence for the harmful effects of dietary sugar. Skeletal remains more than a thousand years old from coastal areas in what is now Georgia (USA) show few dental cavities. They became common with the introduction of maize-based agriculture, and perhaps corn syrup, at about that time. They became still more common with the introduction of other forms of sugar by European settlers.

Cavities are technically not a nutritional problem, but they are a dietary problem and very much a disease of civilization. The good news is that they are of steadily decreasing concern. They were a serious scourge for adolescents and young adults born in the United States before 1940. Advances in preventive dentistry, such as fluoride treatment, have helped to overcome the difficulty, but before these advances could be made it was crucial to realize that sugar is the culprit.

Simple rules and illustrative devices such as Figure 10-1 are always based on conceptual simplifications and all-else-equal assumptions. A diet that is too high in calories and fat for one person may be ideal for another. Much depends on age, size, sex, reproductive processes, genetic factors, and especially activity levels. Early subsistence farmers maintained what might be considered, from an evolutionary perspective, a normal activity level. Except for professional athletes, dancers, cowboys, and a few other groups, most people in modern industrial societies have abnormally low energy expenditures. Workers sitting in swivel chairs or in drivers' seats of cars or even pushing vacuum cleaners or electrically powered lawn mowers are being sedentary, and their leisure hours may be even more so.

During almost all of human evolution, it was adaptive to conserve energy by being as lazy as circumstances permitted. Energy was a vitally needed resource and could not be wasted. Today this take-it-easy adaptation may lead us to watch tennis on television when we would be better off playing it. This can only aggravate the effects of excess nutrition. The average office worker would be much more healthy if he or she spent the day digging clams or harvesting fruit in scattered tall trees. What would an ancestor of a few thousand years ago have thought of the expensive and complicated exercise machine in the office worker's basement—especially if it were actually used?

ADDICTIONS

Historical and anthropological records show that opium and other psychotropic drugs have been available throughout human history, with almost every inhabited region supplying one or more substances with the potential for abuse. Most addicting substances are elaborated by plants as a way of discouraging insect pests and grazers. Many act on the nervous system, and a few just happen to induce pleasure in humans. Alcohol is present in very ripe fruit, and storage of fruit juices yields a beverage with an alcohol content of up to several percent.

Substance abuse today is a greater problem than it was in preindustrial societies because of the technological innovations of the past few centuries or millennia. When every household had to make its own wine or other fermented beverage in small vessels and with primitive equipment, it was unlikely that anyone would have enough for heavy daily consumption. Urban civilizations, with their professional vintners and brewers, were more likely to provide the quantities of alcoholic beverages that would permit the wealthier classes to get all they wanted. Improved methods of storage and transportation, which allowed British tribesmen to get drunk on Roman wines, were another factor in the advance of alcoholism.

Another contribution to this advance was the invention of distillation. The readily available beverages containing a few percent alcohol could then be distilled into ones with high alcohol concentration. It may be easier to succumb to alcoholism by drinking gin than by drinking wine or beer. More recent innovations facilitated the production of heroin from opium and crack from cocaine, concentrates that are more rapidly addictive than the natural substances. The invention of hypodermic syringes is part of the same story. Similarly, the mass production of cigarettes from newly developed tobaccos that caused relatively little throat irritation greatly increased the incidence of nicotine addiction. Despite the great antiquity of addictive possibilities, the modern scourge of substance abuse is largely a product of our abnormal environment.

Of course, as every reader of the headlines knows, addiction is an inherited disorder. We are not sure what the average writer or reader of headlines might understand by this, but what we understand is what we discussed in Chapter 7 as genetic quirks. Some people can have frequent evening cocktails, wine or beer with meals, and occasional weekend binges and never show a sign of alcohol addiction. A person with the relevant genetic quirk will, with the same alcohol intake, show a steady increase in drinking until he or she is spending prodigiously to support an ever-worsening addiction and is ever less able to work and maintain normal social relationships. The consequences of this genetic quirk would have been minimal until after such civilizing inventions as stills and six-packs. Alcoholism and much other substance abuse can justifiably be considered diseases of civilization.

DEVELOPMENTAL PROBLEMS
FROM MODERN ENVIRONMENTS

Lack of adequate exercise may be expected to cause problems other than those associated with overweight and fatty foods. It makes no evolutionary sense, for example, for the human developmental process to cause a large proportion of the population to grow incisors in malfunctional positions and to suffer so many problems with wisdom teeth. If a large proportion of modern children need orthodontia and then later some require expensive and painful surgery on wisdom teeth, it implies that there is something wrong with their environment.

One possibility is a deficient demand for jaw exercise. No Stone Age ten-year-old would have been living on foods of anything like the tenderness and fragility of modern potato chips, hamburgers, and pasta. Their meals would have required far more prolonged and vigorous chewing than is ever demanded of a modern child. We wonder if deficient use of jaw muscles in the early years of life may result in their underdevelopment and indirectly in weaker and smaller associated bone structure. The growth of human teeth is more autonomous, but it assumes a jaw structure of a certain size and shape, one that might not be produced if usage during development is inadequate. Crowded and misplaced incisors and imperfectly erupted wisdom teeth may be diseases of civilization. Perhaps many dental problems would be prevented if prolonged vigorous biting were considered a prestigious athletic attainment for children. Perhaps chewing gum should be encouraged in schools!

Other abnormal behaviors during childhood might cause abnormal physical development. Sitting for hours at a time on chairs or benches in classrooms is unnatural, and nothing of the sort was ever demanded of Stone Age children. When they were sedentary, they would have been squatting, not sitting. Stone Agers must also have been able to shift from squatting to kneeling to walking or running or other sorts of activity. Might it not be that many of today's sufferers from lower back pain owe their distress to the hours of abnormal posture imposed day after day during childhood? Maybe the later problems could be avoided by having children do more squatting and less sitting and giving them more exercise breaks or walks between classes.

University of Michigan physician Alan Weder and his colleague Nicholas Schork have tried to understand high blood pressure as a disease of civilization. Instead of emphasizing the high levels of salt in our diets, however, they note that blood pressure must be higher to supply the needs of larger bodies and that there is a mechanism that increases the pressure during adolescent growth spurts. In the ancestral environment, they argue, this mechanism would have made adjustments within a range of small body sizes. Today, our nutritionally rich environment yields fast growth and large body sizes that were rare in the past. The blood-pressure-regulating mechanism, pushed to adjust the system outside the range for which it was designed, often overshoots, causing high blood pressure.

Myopia is not the only ocular abnormality that may arise from novel environmental conditions early in life. Medical science has only recently become aware of ways in which eye usage in the first weeks and months after birth may be critical to the normal development of vision. Preferential use of one eye rather than the other, from whatever cause, may lead to changes in the allocation of brain regions to ocular functions so that a child may later prove incapable of using binocular cues for depth perception. Twenty-four-hour bright lights, sometimes used to treat neonatal jaundice, can cause color-vision defects not likely to be detected until much later. Would it be surprising to discover that constant exposure to loud noises, especially the unchanging sounds of modern machinery, can cause defective hearing development in some babies?

OTHER DISEASES RESULTING FROM MODERN ENVIRONMENTS

Cold weather can be considered a novel environmental factor. The spread of human populations to seasonally cold environments was facilitated by technological innovations, such as clothing and fire, which we achieved only a few tens of thousands of years ago. We still need these artificialities, or their modern equivalents, to survive the winter over much of the currently inhabited surface of the earth. Technology compensates for human biological inadequacies in dealing with such novel environmental threats as frostbite and hypothermia.

But low temperature is not the only stress imposed by high latitudes. Clothing and shelter that enable us to survive in places like Montreal and Moscow impose their own health problems. Our synthesis of vitamin D depends on our exposing our skins to sunlight. If we are indoors much of each day and largely covered with clothes when we are out, the amount of vitamin D we synthesize will be a tiny fraction of that made by a naked forager on the African savannah, and it could be grossly inadequate for our metabolic needs. Fortunately, our photosynthetic capability is not our only source of this material. We can also fulfill our vitamin needs by eating certain foods. Unfortunately, a seemingly adequate diet may in fact provide very little vitamin D, and a deficiency leads to health problems related mainly to abnormalities of calcium metabolism.

The most commonly recognized effect of vitamin D deficiency is rickets, a developmental disease of childhood. The symptoms are many, but the most important is defective growth of the bones. They become soft and weak from deficient calcium deposition and grow abnormally. The disease is essentially unknown in the tropics, where everyone gets abundant sunshine, and uncommon in Japan, Scandinavia, and other regions where traditional diets include good sources of vitamin D, such as fish. But at times it affected such large numbers of children in England that it was sometimes called the English disease.

Rickets was also a frequent malady in northern American cities prior to the 1930s, when vitamin D began to be routinely added to milk. Rickets struck black children at a higher rate than white. The adaptive significance of human racial differences is generally dubious, but the reduced vulnerability of pale-skinned people to rickets may be a valid example. Perhaps the first people who crossed the Mediterranean and later the Alps were quite dark. They found a land covered with trees under a sky often covered with clouds. During much of the year they spent long hours huddling in caves or drafty shelters. When they went outdoors they clothed themselves with animal skins or woven fabrics and exposed very little skin to the meager sunshine. The result, for many people, may have been depressed fitness because of vitamin D deficiency. Those who happened to have less heavily pigmented skins, which admitted more light for vitamin D synthesis, would have fared better than their darker neighbors.

In this way light skin may have evolved in perhaps a few hundred generations. The change may have been rapid because reductions of a trait are generally easier to evolve than increases or elaborations. Cave animals may lose almost all ability to make pigment in a few thousand generations, and this happens merely from relaxed selection for the maintenance of color. If there is an actual advantage to paler skin, the change should be much faster. The same evolutionary reduction of melanin synthesis may have happened, though to a lesser extent, in the colder parts of Asia, where forests give way to grasslands and deserts and winter days are more often sunny. The native peoples of Siberia and northern China are darker than those of central and northern Europe but paler than those of Africa or southern Asia. As a disease of civilization, rickets is more of a hazard for people with highly pigmented skin, and pale skin may be recognized as especially adapted to a scarcity of sunshine. But then what happens when these pale people move back to sunny regions, such as Australia? Stay tuned for more on the sunshine problem (Chapter 12), and recall our discussion of sunburn in Chapter 5.

As noted above, the invention of agriculture led to population densities much greater than could be achieved by hunter-gatherer economies, and it permitted the support of great concentrations of people in cities. The spread of people into seasonally cold environments led to their prolonged concentration inside caves and buildings. Both these changes increased the number of people a given individual would contact in a short period of time and increased the closeness and duration of such contacts. New infectious diseases could then emerge that could be spread only by abundant personal contact.

Much of the natural selection taking place in these populations may have consisted of the weeding out of individuals whose genetic quirks made them vulnerable to smallpox, measles, or other contact-transmitted diseases. High-cost defenses against such tropical diseases as malaria, for example the sickle-cell trait, would have been lost rapidly. The effectiveness of the newly evolved defenses against such diseases as smallpox was tragically shown when settlers, carrying what for them were well-controlled pathogens, invaded parts of the world where native peoples had never been exposed to the diseases of civilization. Far more New World people were killed by European diseases such as smallpox and influenza than by European weapons.

In this chapter we have scarcely hinted at the many psychological problems that may arise from modern life. Despite the family-values rhetoric of politicians, children raised by nuclear families in single-unit suburban dwellings are experiencing a profoundly novel social environment, as are those being supervised by transient caretakers in day care centers. As adults and even as adolescents and children, we may have to deal more often with impersonal bureaucracies than with familiar individuals. Most of the people we encounter on what seems to be a normal day may be strangers. This was not the kind of world our ancestors evolved in. What about the prolonged winter darkness of high latitudes and, conversely, the hours of bright indoor lighting and resulting shortened periods of real darkness we actually experience? The cabin fever of snowbound Alaskan gold seekers is now a recognized malady that is getting attention from medical researchers. What about night-shift workers and the jet-lagged jet set? And then there are the psychological—as well as physiological—effects of offices without windows. We have just begun to explore the medical consequences of our novel modern environment.

CONCLUSIONS AND RECOMMENDATIONS

There is no Eden we can go back to even if such a move were desirable. What we can do is be alert to the modern dangers and take reasonable steps to forestall them. As with many other topics discussed in this book, our main recommendation for anyone faced with a problem of medical importance is to consider the question: What is its evolutionary significance? One possibility is that it is an adaptive mechanism, but this will normally mean adaptive in the Stone Age. Our cravings for sugar and fat, our tendencies to be lazy, and our eye-growth adjustments that result in myopia are evolved adaptations, but in modern environments they cause difficulties for many people. Other evolved attributes, such as senescence and susceptibility to sunburn, are adaptive in no environment but may represent costs of other adaptations. Again and again we harp on the themes that all benefits have costs and that many benefits are worth their associated costs.

11

ALLERGY

Many people in temperate parts of North America dread the day in August when ragweed first releases its pollen, causing sneezing and wheezing and reaching for handkerchiefs and antihistamines. The poor ragweed plant is just trying to reproduce, but we are the ones who suffer. A single plant may release a million grains of pollen a day, mostly between 6 and 8 A.M., perfect timing to maximize the likelihood that those sex cells will find their way to receptive ragweed flowers on the morning breeze. A square mile of ragweed plants can produce sixteen tons of pollen in a year, but an allergic response can be provoked by one millionth of a gram. The notorious pollen grain is tiny, a sphere twenty microns in diameter that contains two living ragweed sex cells, accompanied by proteins and other nutrients. One of the proteins, Amb a I, makes up only 6 percent of the protein but causes 90 percent of the allergic activity. And what a lot of unfortunate activity it is. From the middle of August, those who suffer from ragweed allergy look forward to the day a few weeks before the first hard freeze when the ragweed plants will die and stop broadcasting their pollen.

Ragweed is, of course, not the only culprit. Allergies are also provoked by inhaling other pollens, fungal spores, animal danders, and mite feces, by skin contact with many different substances, by eating certain foods or drugs, and by injections of drugs or toxins like bee venom. A quarter of the modern American population suffers from some allergy or another. You or a relative or friend may well have sought help from an allergist. If so, you probably had skin tests to try to identify the substance (*allergen*) that caused the allergy. Two kinds of advice were then forthcoming: avoid the allergen and relieve the symptoms with this or that anthistaminic drug.

Avoiding the allergen makes sense, but what about relieving the symptoms? We dealt with that kind of advice in discussing the treatment of infectious disease. Could taking antihistamine for your allergy be analogous to taking acetaminophen for fever or giving mice a pill to keep them from smelling cats? At the moment we know that the system that gives rise to allergy is a defense, but we do not know for sure what it is supposed to defend us against. We can be sure that the capacity for an allergic reaction is a defense against some kind of danger, or else the underlying mechanism, the immunoglobulin-E (IgE) part of the immune system, would not exist. It is perhaps conceivable that our IgE system is a remnant of a system that was useful for other species, but this is unlikely because systems of this complexity degenerate quickly if they are not maintained by natural selection and even more quickly if they cause any harm. It is much more likely that the IgE system is somehow useful.

This need not mean that every allergy attack is useful. In fact, an evolutionary view of inexpensive defensive reactions suggests that most individual instances will be harmful even though the system as a whole is adaptive. This is a manifestation of the *smoke-detector principle*. Smoke detectors are designed to warn people when a dangerous fire is in progress, but few of them ever perform this service. They hang there year after year doing nothing or only sounding an occasional false alarm from a cigar or smoky toaster. Yet the annoying false alarms, and the costs of the smoke detector and its occasional battery change, are well justified by the protection they provide against a major fire. More on this principle when we discuss anxiety in Chapter 14.

Your allergist probably did not give you a discussion about the utility of the IgE system and the evolution of its regulation. If you asked why you have to be allergic to cats or oysters or whatever, your allergist probably said something like "Well, as in everything else, people vary tremendously in their sensitivities to different allergens, and you happen to be excessively sensitive to something in cat dander. This excess in your sensitivity must be treated by avoiding cat dander and suppressing the defensive reaction it triggers."

There are two serious difficulties with the excess-sensitivity theory. First, an allergy is not just a matter of degree. Allergic people react to minute traces of their allergens, while nonallergic people have no apparent reaction to enormously greater quantities. In this respect allergy is quite different from an excess sensitivity to sunshine or motion sickness. The second difficulty is more serious. Allergy is not an extreme action of some normally well behaved system with an obvious function. IgE antibody seems to do almost nothing, at least in modern industrial countries, except cause allergy. It would appear that we evolved this special IgE machinery for no better reason than to punish random individuals for eating cranberries or wearing wool or inhaling during August.

Despite these problems, this explanation of allergies as a result of excess sensitivity is widely employed. For instance, a 1993 *New York Times* article on asthma describes it as an excessive immunological reaction, one to be solved by finding a drug that can "interfere with the asthmatic process" by "keeping the lungs from responding to allergens in the first place." Nowhere is the possibility considered that the lungs (or their IgE-carrying cells) may know something that we don't. A widely used textbook of immunology describes allergy in a chapter entitled "Hypersensitivity" and also makes no effort to explain why the IgE cells exist at all.

THE MYSTERY OF THE IGE SYSTEM

On finding a complicated feature characteristic of a species or larger group, one of the first things biologists want to know is what it does. They assume that if it did not do something important it would not have been produced and maintained in evolution. A short digression offers a vivid illustration. The snouts of sharks contain a cluster of flask-shaped organs (the ampullae of Lorenzini, named for the Renaissance anatomist who first described them). These complicated structures have a rich nerve supply. For three centuries people guessed that the ampullae of Lorenzini regulated buoyancy or amplified sounds, but no serious biologist suggested that they were "just there." The question stayed on the table until some adequate experiments finally showed that the ampullae of Lorenzini detect minute electrical stimuli, thus allowing sharks to detect muscle activity in potential prey hidden in total darkness or buried in the sand. This discovery was made only because some biologists, habitués of the adaptationist program, assumed that the ampullae of Lorenzini must be an adaptation.

Before we discuss possible explanations for the IgE system and the allergies it causes, we need to describe the proximate mechanisms of allergy. When a foreign substance enters the body, it is taken into cells called macrophages (macro means "big" and phage means "to eat"), which process the proteins from the substance and then pass them on to white blood cells called helper T cells, which take the proteins to another kind of white cell called B cells. If the B cell happens to make antibodies to that foreign protein, it is stimulated by the T cell to divide and make those antibodies. Most often that antibody is the familiar immunoglobulin G (IgG), but, for certain substances, the B cell is instead induced to make IgE antibody, the substances that mediate allergic reactions.

There is remarkably little IgE, compared to other antibodies. It makes up only one hundred-thousandth of the total amount of antibody. The IgE antibody circulates in the blood, where about one out of one hundred to one out of four thousand molecules attaches to the membranes of still other cells called basophils (if they are in the circulation) or mast cells (if they are localized). When attached to these cells, the IgE remains for about six weeks. Despite the small amount of IgE, there will still be between 100,000 and 500,000 IgE molecules on each basophil, and, in an individual allergic to ragweed, about 10 percent of IgE may be specific to ragweed antigens.

These mast cells are primed, like mines floating in a harbor, waiting for reexposure to the allergen. When it does return and is bound by two or more IgE molecules on the surface of the mast cell, the cell pours out a cocktail of at least ten chemicals in the space of eight minutes. Some are enzymes that attack any nearby cells, some activate platelets, some attract other white cells to the site, while others may stimulate smooth muscle (causing asthma). One, histamine, causes itching and increased permeability of membranes, unpleasant effects that can be blocked by antihistaminic drugs. While the details are still being worked out, the general operations of this proximate mechanism have been known for about twenty-five years and are essentially the same in all mammals.

At this point you may be thinking: surely by now someone must have figured out what all that IgE machinery is there for! People have tried, but so far there has not been enough serious research to arrive at a generally accepted explanation. Many thoughtful researchers are well aware that a system this sophisticated must have some useful function. "These cells are not simply troublemakers devoid of redeeming biological value," says Stephen Galli from Harvard, who notes that the distribution of mast cells adjacent to blood vessels in the skin and respiratory tract places them "near parasites and other pathogens as well as near environmental antigens that come in contact with the skin or mucosal surfaces." Galli does not, however, review evidence about the possible functions of the system. A new nine-hundred-page textbook on allergy devotes only one page to the problem. It notes that "Several roles for the possible beneficial effect of IgE antibody have been postulated," including regulation of microcirculation or as a "sentinel first line of defense" against "bacterial and viral invasion" and attacking parasitic worms. It concludes, "With 25% of the population having significant allergic disease mediated by the IgE antibody, an offsetting survival advantage for the presence of IgE has been suggested." But, like other textbooks, it never seriously tries to explain the adaptive significance of allergy.

The most widely accepted view is that the IgE system is there to fight parasitic worms. Evidence for this idea comes from the observation that substances released by worms may stimulate local IgE production and the resulting inflammation, which are interpreted as defensive activities against the worms. Further evidence comes from experimental studies of rats that developed strong IgE responses to *Schistosoma mansoni* infections. Transfer of IgE from one rat to another transfered protection against infection, while blocking the ability of IgE to recruit other cells made the rat more vulnerable to the worms. In people infected with schistosomes, 8 to 20 percent of their IgE may attack these worms, and those with a decreased ability to make IgE have more severe infections.

Worms such as schistosomes, which cause liver and kidney failure, and filaria, which cause blindness, were all substantially greater problems before the introduction of modern sanitation and vector control. If attacking worms is the only function of the IgE system, this supports the current practice of treating allergies in developed countries by inhibiting allergic symptoms because an allergic reaction to anything but a worm would be maladaptive. However, the evidence that attacking worms is the only or even a major function of the IgE system remains inconclusive, and some of it may be flawed by attempts to interpret the data in terms of the only available hypothesis. Alternative explanations for the association of IgE phenomena with worms, such as the possibility that worms arouse IgE responses for their own benefit (by increasing the local blood supply), have been insufficiently considered.

There is, however, another possible function for the IgE system, one recently championed by Margie Profet, whom we met in our chapters on signs and symptoms and on toxins. Profet proposes that the IgE system evolved as a backup defense against toxins. As we argued in Chapter 6, our environment is and always has been full of toxins. Inhaled pollen, contacted leaves, and ingested plant and animal products all contain potentially harmful substances. Most of these toxins are formed by plants to protect themselves against parasites and insects or other plant-eating animals.

We have several kinds of defenses against these chemicals. First, we avoid them when we can. Also, the linings of our respiratory and digestive systems are equipped with toxin-fixing antibodies of the IgA group and with detoxification enzymes that collectively decompose broad categories of chemical structures. Mechanical defenses provided by mucous secretions and by the structure of our skin and absorptive surfaces also play a role. Toxins that bypass these initial defenses are attacked by concentrated batteries of enzymes in our liver and kidneys.

But suppose all these defenses fail, as all adaptations must sometimes. Then, according to Profet, comes the backup defense, allergy, which gets toxins out of you in a hurry. Shedding tears gets them out of the eyes. Mucous secretions and sneezing and coughing get them out of the respiratory tract. Vomiting gets them out of the stomach. Diarrhea gets them out of parts of the digestive system beyond the stomach. Allergic reactions act quickly to expel offending materials. This fits with the rapidity with which toxins can cause harm. A few mouthfuls of those beautiful foxgloves in your garden can kill you a lot faster than a phone call can summon first aid. Appropriately for Profet's theory, the only part of our immunological system that seems to be in a great hurry is that which mediates allergy. Other aspects of allergy that she mentions in support of her theory include the propensity to be triggered by venoms and by toxins that bind permanently to body tissues, the release of anticoagulants during allergic inflammation to counteract coagulant venoms, and the apparently erratic distribution of allergies to specific substances.

At this point we pause to line up our ducks in a row so we can aim at them, even though we don't yet have a way to shoot them. As we have already noted, the first and most important question is, What are the normal functions of the IgE system? The second question is why some people are especially susceptible to allergies while others are not. The third question is why a susceptible person develops an allergy to one substance and not another, say, milk instead of pollen. The fourth question is why allergy rates seem to be rapidly increasing in recent years.

ATOPY

People who are especially susceptible to allergies are said to be "atopic." Atopy runs strongly in families. While the risk of clinically significant allergy in the general population is about 10 percent, the risk is closer to 25 percent if you have one atopic parent and 50 percent if both your parents are atopic. The responsible genes remain elusive, but a dominant gene on chromosome 11 may play a key role. If the genes that predispose to allergy are found, we will still need to find out why they exist. Do they, like the sickle-cell gene, give an advantage in certain environments or protection against certain infections? Or do they give an advantage when combined with certain other genes but a disadvantage otherwise? Or might they be "quirks" that did not cause disease until they interacted with modern environments?

Genes are not the whole story, though. Studies of identical twin pairs show that in half the cases, one twin has allergies while the other twin is unaffected. So factors other than genes must be important as well. And even among atopic individuals, one may be allergic to ragweed while the other is allergic to shrimp. Why? As a start toward answering this question, we will invoke two ideas, one being the tendency, discussed above, for defensive adaptations to make many of the cheap kind of mistake in preference to the expensive kind (the smoke-detector principle). The other derives from the phenomenon of enzymatic variability, which has gotten considerable recognition in the recent biological literature.

Specimens of the same species, human or otherwise, can be immensely variable. Their genetic codes may be 99 percent identical, but tiny differences in genetic code can result in strikingly different structures and body chemistry. The parts of the code that are the same can also code for differences, if they include instructions of the form "if A then X, else Y." In retrospect we see that the evidence for wide variation among individuals has always been there. Just consider how different males and females of many species can be in size and anatomy, reproductive processes, behavior, and often in diet, habitat, and other features. These differences may result from genes that are expressed only if testosterone above some threshold concentration is present. The best examples of human variations are differences in reactions to drugs. Some individuals may take ten times as long as others to reduce a drug concentration to half its initial value. To put this into perspective, suppose you and your friend each get the same injection of quinine; it takes you an hour to detoxify half of it, and his system does this ten times as fast. At the end of the hour, when your concentration is still half what it was initially, his is down to less than a thousandth of its starting value. If the enzyme is cholinesterase and the drug is a cholinesterase inhibitor, often used to relax muscles during surgery, such slow metabolism might leave you still paralyzed and unable to breathe hours after other patients have been up and around. Anesthesiologists are, thankfully, on the lookout for individuals with this idiosyncrasy.

If Profet's theory is right, people may develop allergies to the specific toxins to which they are especially vulnerable. Consider President Clinton, who is allergic to cats. Could it be that this allergy protects him from some dangerous toxin? Remember that the pitohui bird (Chapter 6) has toxic feathers. It seems unlikely that cats have a comparable adaptation, but let's consider the possibility. Why should Bill Clinton be vulnerable when none of his relatives are? Perhaps merely because he inherited defective forms of some gene that makes an enzyme important in denaturing some cat toxin. If he touches cat fur or inhales microscopic particles of it, the toxin would enter his cells and reach harmful concentrations, instead of being quickly destroyed by the enzymes normally present. Fortunately, the president has mast cells and IgE-producing T cells that react to the toxin by triggering defensive reactions, such as sneezing. This might mean that he has to interrupt important negotiations to yank a handkerchief out of his pocket, but the sneeze, as a backup defense, might save him from some serious malady. Do you believe this explanation for Bill Clinton's allergy to cats? We don't, but we have a good excuse for telling it. At the moment, there is no evidence that it is wrong. As long as we do not know what the IgE system is for, we will have great difficulty distinguishing its accomplishments from its mistakes.

We can alter the story to make the cat allergy a nuisance without value, while still basing the explanation on Profet's theory of allergy as a backup defense against toxins. Perhaps Bill Clinton's allergy is just another example of the smoke-detector principle. Perhaps as a child he encountered bacterial toxins during a respiratory infection, and his IgE system went into action and reacted, not only to the dangerous material, but also to some innocent "bystander" molecules (Profet's term). Perhaps some harmless component of cat fur was mistakenly perceived, by a few IgE-producing cells, to be a troublesome toxin, or at least a reliable sign of the toxin's presence. Immune cells that react to a foreign substance multiply and become far more numerous. So after this first episode, large numbers of anti-cat cells were poised to go into action on the next exposure. Do you prefer this explanation for Bill Clinton's allergy? We do, but we are not inclined to bet on it. There is just not enough information for an informed decision.

If you were the president's physician, what would you recommend? Would you prescribe a drug to inhibit the allergic reaction? The answer should depend on whether the allergy is useful or not. Is it an effective defense against an otherwise dangerous toxin, or is it a false alarm? How do you decide? At the moment, you have no solid basis for deciding. You might want to use antihistaminic drugs to suppress the allergic reaction, since they have no known dangers, but there are no adequate antihistamine studies that would detect the kinds of dangers implied by Profet's theory.

The possibility of harm resulting from suppressing the symptoms of allergy is of special concern because of data suggesting that allergy may protect against cancer. Profet reports that sixteen out of twenty-two epidemiological studies found that people with allergies are less likely to have cancers, especially of tissues that show allergic reactions. On the other hand, three of the studies found no clear relationship, and three others, including one large, well-controlled investigation, found that some allergies are associated with an increased likelihood of developing some cancers. What are we to make of this? It would certainly be premature to conclude that allergies protect against cancer, but it is not premature to begin looking at the possible risks of long-term use of medications that suppress allergic responses. Unfortunately, the nonmedication treatments are mainly inconvenient or not very effective. If you've got hay fever, you may be hard put to follow your doctor's advice to stay indoors in closed rooms as much as possible, wear a pollen mask when you must be outdoors, or go somewhere else for the bad season. Taking a pill is much more convenient.

If the antitoxin theory of allergy is correct, it has clear implications for medical research. A Utopian recommendation is simple: find out just what the toxins are in pollen, cats, seafood, and so on, that induce allergy and devise techniques for their denaturation. These toxins may be different from the antigens that stimulate the allergy. If we knew just what was dangerous about ragweed pollen, we could perhaps equip people with nose drops or inhalants that would chemically inactivate both the toxin and the antigen. We could treat allergenic foods in similar ways. If we knew which patients don't need their allergies to compensate for some deficiency in their ability to detoxify, we could suppress their symptoms without concern.

Such studies will be inconclusive unless they can distinguish useful allergies from useless ones. If Profet is correct in reasoning that an allergy to eggs is consistently maladaptive, this allergy should not protect against cancers of the digestive tract, and the inflammation caused by the allergy might even increase the risk of cancer. An allergy to shrimp, however, would be expected to decrease the cancer risk for anyone who is unable to detoxify one of the many noxious compounds that shrimp get from their phytoplankton diets. Profet's theory provides a basis for predicting when allergy will protect against cancer and when it might be irrelevant or actually increase the risk. We should emphasize that her theory is novel. Few allergists have even heard about it; far more believe the antiworm theory. But either theory may be better than no theory at all. As Thomas Huxley once observed, truth is more likely to emerge from error than from vagueness.

Still another possible function of the IgE system may be to defend against ectoparasites such as ticks, chiggers, scabies, lice, fleas, and bedbugs. A small problem for most people in modern societies, ectoparasites have been, throughout most of human evolution, not only a constant nuisance but vectors for many diseases. Slapping, scratching, and mutual grooming are only partially effective defenses. When cows are prevented from grooming by a thick collar, their burden of ticks and lice increases steadily and then suddenly crashes when the cow's immune systems begin responding to a bite with an inflammatory response that makes it impossible for the parasites to get a blood meal. Prevention of ectoparasite infestation might explain many aspects of the IgE system, especially the concentration of mast cells on the body's surfaces, the immediate massive response, and the stimulation of itching. This theory could be tested by looking to see if the immune response that counters ticks on cows is indeed based on IgE and by looking at the IgE responses of people who are infested with ectoparasites.

As with other traits, the IgE system may well have more than one function. Some combination of the above and other explanations may be correct. One of the best ways to determine the function of a trait is to observe the problems of those who lack it. The deficits of a person who lacks eyes are obvious, and those of a person without kidneys soon become apparent, but the functions of many traits are more subtle. The spleen, for instance, is usually surgically removed if it ruptures, as it sometimes does in automobile accidents. Such patients have no apparent disability, but if they are stricken with pneumonia, the infection may quickly kill them because the spleen is not there to filter infectious particles out of the blood.

What happens to people who lack the ability to make normal IgE? While some people with very low levels of IgE are healthy, others are plagued with recurrent infections of the lungs and sinuses as well as fibrosis of the lungs. While these findings could be a result of exposure to toxins or a secondary result of whatever factor caused the IgE deficiency, there is also evidence for specific IgE antibodies directed against *Staphylococcus aureus* in people who cannot make other immunoglobulins. In a study of 190 patients with bronchial asthma, 55 had IgE antibodies to substances in the bacteria *Streptococcus pneumoniae* and/or *Haemophilus influenzae*. Furthermore, one effect of the substances released by mast cells is to attract other immune defense cells to the area, where they are available to fight any invader. All this suggests that the IgE system may directly or indirectly defend us against ordinary bacteria and viruses. The complexity of the immune systems, with functions that overlap and back one another up, makes it difficult to identify the benefits of the IgE system. It will take patient, well-designed research to answer the important but unanswered question, *What is the IgE system for?*

The Most Worrisome Question

Another puzzling aspect of allergy, at least respiratory allergy, is the apparent recency of its appearance as a major medical problem. John Bostock originally described his own symptoms of hay fever for the Royal Society in 1819 and later reported that he could find only twenty-eight cases after investigating five thousand patients in all of England. Records imply that hay fever was essentially unknown before 1830 in Britain and 1850 in North America. In Japan its incidence was negligible in 1950, but it now affects about a tenth of the population. If the increase is real and not just an artifact of inadequate records, what novel environmental factor of the last century or two can account for this alarming phenomenon?

One clue comes from studies of the factors that seem to sensitize predisposed individuals, mainly exposure to antigens in the first two years of life. In one study of 120 infants with high susceptibility to allergy on the basis of their IgE levels at birth, 62 were raised as a control group without any intervention, while the mothers of 58 in the experimental group were taught how to keep their homes relatively clean of allergens, prevent mites, and avoid giving potentially allergenic foods to their infants. At age ten months, 40 percent of the control group had developed allergies compared to only 13 percent of the experimental group. Perhaps part of the increasing rate of allergy results from living indoors with drapes and wall-to-wall carpets, which provide breeding places for dust mites.

When Eric Ottesen, head of the Clinical Parasitology Section at the National Institute of Allergy and Infectious Disease, studied the six hundred people who live on Mauke, an atoll in the South Pacific in 1973, only 3 percent of them had allergies. By 1992, the rate was up to 15 percent. He suggests that institution of treatment for worm infestations during the intervening years left the IgE system with no natural target, so that the usual mechanisms that downregulate the system are inactive and the IgE begins to attack harmless antigens.

Breast-feeding decreases the incidence of allergies, so bottle-feeding may also contribute to the rise in allergies. Perhaps babies deprived of maternal antibodies make more immunological mistakes in coping with antigens on their own. Or perhaps crowded, mobile modern societies expose infants to a greater diversity of viral respiratory diseases and thereby greater exposure to miscellaneous allergens. The increased quantity and variety of atmospheric pollutants may foster increases in both helpful allergies (if such there be) and harmful ones, perhaps because chemical damage to the respiratory mucosa may admit antigens that would otherwise be kept out. Food allergies, although perhaps not as clearly on the increase, may have become more troublesome because we now have so little control over what we are really eating. Eggs, wheat, soybeans, and other possible allergens may be present in a great variety of commercially prepared foods and be extremely difficult to avoid, even by people who know they are allergic to them.

What are we doing today that is different from what we did just a century ago and that makes us so much more vulnerable to so many diverse allergies? We desperately need real answers. Respiratory allergies affected less than 1 percent of people in industrial societies in 1840. Now, a hundred and fifty years later, it afflicts 10 percent. What might the future hold if we remain as ignorant as we are now?

12

CANCER

O n March 5, 1992, *The New York Times* carried an obituary for well-known actress Sandy Dennis, a cancer victim at fifty-four. That same day, the eighty-three-year-old actress Katharine Hepburn was enjoying her autobiography's twenty-fifth week on the *Times*'s best-seller list. An obvious question is, Why did cancer strike Sandy Dennis? What caused her to miss out on the long life that her fellow actress enjoyed?

This obvious question is morally and medically a good one, but there is a more profound biological question: How is it possible that any of us can live several decades without dying of cancer? Cancerous cells are merely cells doing their normal thing: growing and proliferating. How could so many cells do such an *abnormal* thing as inhibit their growth for many decades? Obviously they must; otherwise everyone would die of cancer at an early age. This, of course, is the ultimate explanation. Those least likely to die at an early age, from any cause, will be most likely to survive, reproduce, and have their cancer-delaying adaptations at work in future generations. This sort of evolutionary explanation can help us understand the workings and origins of our cancer-preventing adaptations and the prodigious accomplishment they represent.

Confucius once said something like: A common man marvels at uncommon things; a wise man marvels at the commonplace. To marvel at the commonplace of not having cancer and at the mechanisms that make this possible may be the key to understanding how to make cancer even more uncommon.

THE PROBLEM

The magnitude of the problem of avoiding cancer may be appreciated from considering the long-term history of any cell in our bodies. A cell now contributing to normal functioning in the liver of some Hollywood star arose by the growth and division of some preexisting cell, probably one closely similar to itself. That parental cell arose from another before that, and so on. As we trace the ancestry of the liver cell, we find cells that look ever less like liver cells and ever more like undifferentiated embryonic cells. Some years back in the cell lineage we come to the fertilized egg from which the entire individual arose.

That cell had a history too, a lineage through various oocytes and oogonia back to the embryonic cells that developed into the Hollywood star's mother. Likewise, the sperm that did the fertilizing came from a lineage of spermatocytes and spermatogonia back into the embryonic cells of our star's father. Thus back through the mother's and father's original zygotes into the grandparental generation, and so on in endless repetition of ever-dividing embryonic and reproductive cells. Never in these sequences of cell divisions, for the billion years or so since the origin of the first real cells, was there ever one that did not divide, and nowhere in these lineages was there anything that looked like a liver cell.

We offer Figure 12-1 as an aid in understanding this essential fact of life. All our ancestors had livers, but none of the cells of these ancestral livers gave rise to any of our liver cells, or to anything else in our bodies. We arose entirely from a line of endlessly proliferating germ-line cells. This picture, of an eternal *germ plasm* giving rise to elaborate somata of individuals, which are always genealogical dead ends, was first presented by August Weismann, a nineteenth-century Darwinian.

Now, for the first time in these eternal lines of descent and after dozens of the cell divisions needed to create an adult soma from a single cell, we find a cell, say a liver cell, that must play a specialized role in the life of a multicellular individual. This liver cell must do something none of its ancestors ever attempted: it must stop dividing. If there is an injury to the liver, the cell may be called upon to divide again. This sort of growth and division must be in precisely the

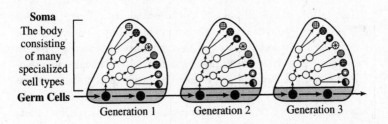

Soma
The body
consisting
of many
specialized
cell types
Germ Cells

Generation 1 Generation 2 Generation 3

FIGURE 12–1. Germ plasm concept of Weismann. The eternal line of germ cells gives rise to individual bodies with a limited life span. The individuals diagrammed can be of either sex.

amount and pattern required for normal liver function and must cease as soon as this machinery is fully restored. If ever, in any one of the billions of cells of the liver, the growth and division process is turned on inadvertently and proceeds unchecked, a tumor develops and eventually causes a lethal disruption of some physiological function.

From this perspective, life seems rather precarious. It suggests that we must have some really superb anticancer mechanisms acting in our favor. As American marine biologist George Liles observed, "the cells and organs that make life possible had better be well designed, because the job of living is formidable. Living beings—plants and animals, bacteria and slime molds and fungi—every animate entity faces a set of challenges that would give pause to the most inventive designer." He was led to this remark in considering what might seem a rather simple sort of problem, the proper routing of water through the feeding machinery of a mussel. How much more formidable is the challenge of avoiding cancer for several decades in the collection of ten trillion cells that make up a human being!

Biologists today more or less universally believe that multicellular organisms, such as ourselves, arose from some group of the protozoa, in which each cell was a functionally independent individual. Most of their reproduction was asexual, with one cell dividing to form two new ones. In some modern protozoan species these two new individuals do not break completely apart but stick together in pairs. In others, the offspring of pairs stick together in filaments or sheets called colonies. In a few, the colonies may differentiate into germ cells and somatic cells, as shown in Figure 12-1. This means that some previously independent cells, apparently voluntarily, give up reproduction and become genealogical dead ends. They devote themselves entirely to supplying nutrients and protection to the few germ cells that ultimately participate in sexual reproduction. Some such sequence of developmental events, as observed in the much-studied colonial protozoan *Volvox carteri*, must have characterized some remote ancestor of all multicellular animals.

Can this acceptance of a sterile, servile role be explained by natural selection? The answer is obviously *no*, if this process means selection among cells for those best able to survive and reproduce. The answer is *yes* if the selection is among the genes best able to get themselves into future generations. If the reproductive and somatic cells of a *Volvox* colony have the same genes, it does not matter which cells actually do the reproducing and which become sterile. All that matters is that the sterile cells, in their strictly somatic roles, make the colony's reproduction of genes identical to their own more successful than if they too formed eggs or sperm. If colonies with ten reproductive cells and a hundred sterile ones reproduce more successfully than those with eleven and ninety-nine, the tendency for most of the colony cells to assume a somatic service role will be perpetuated.

A colony of a hundred cells, all derived in a short time from a single original cell, may well be all of about equal health and vigor and will almost certainly be of the same genotype. The resources needed to produce a hundred cells from one may all be shared equally, and all cells have elaborate mechanisms for protecting the genetic material from damage or alteration. But what about a thousand or ten thousand cells? Would colonies that big be asking for trouble? Might there not be occasional mutations that would make cells behave in ways other than those that maximally benefit the colony as a whole? For instance, might not such a mutant cell start appropriating more than its maintenance requirements for nutrients and start growing and reproducing, even though this might be harmful for the colony? Such large colonies surely need special adaptations for maintaining discipline among the many component cells.

THE SOLUTION

How about a colony the size of an adult human body? What sort of special adaptation would be adequate to maintain discipline among ten trillion cells? From an engineering perspective, it is difficult to imagine how any quality control system would be equal to the job. An auto manufacturer faced with turning out a mere ten thousand vehicles, not one of them with a serious flaw, would be well advised to quit the business. A single living cell is incomparably more complicated than any automobile.

Consider the problem faced by an embryo of a hundred cells that gives rise to one of a thousand that produces one of ten thousand and so on to the ten-trillion-cell adult. Most of these cells will die and be replaced by others. All these cells are equipped with genes that turn out products essential to their division, and some genes are adjusted so as to stop making this product when local conditions indicate that the tissue is mature and no additional cells are currently needed. If one of these genes gets accidentally altered in a way that makes it heedless of these conditions and the gene goes on making its product, mechanisms of DNA editing and repair step in and correct the flaw—or at least they are supposed to. One out of about two hundred people has a gene that greatly increases the likelihood of colon cancer. Originally thought to be a gene that actually did something to cause cancer, it is now recognized as a defective form of a normal gene that acts in the detection and rectification of abnormal DNA structure. When this system is not working, DNA abnormalities accumulate and the chance of cancer increases drastically.

Very few such flaws actually get a chance to express themselves. How few? Let's assume that only one such gene in ten thousand cells makes its product when it is not supposed to. Starting with ten trillion cells, we can assume there are a billion altered cells, scattered through the body, that are capable of initiating a cancerous growth. This is not all that reassuring. But there is another kind of genetic safeguard in each cell: tumor-suppressor genes that actively inhibit cell growth, perhaps by destroying the product of a gene that makes a substance essential to division, when it is inappropriate. Let's assume that this safeguard is also fantastically effective and that the daily rate of failure is only one in ten thousand cells. We can now assume we have only a hundred thousand cancers beginning in the body each day. If there were three equally reliable safeguards and abnormal cell division could not begin unless all three failed, there would still be ten new cancer cells formed each day. This is still not very reassuring.

The situation is analogous to the problem of command and control of nuclear missiles. The risk of catastrophe from accidental firing is so great that the system is designed first and foremost to prevent accidental firing, even at the risk of sometimes not being able to fire when needed. This is the exact opposite of the smoke-detector principle we described for defensive responses. Control of cell division could be said to be based on the principle of "multiple safety catches." The crew in the missile silo cannot fire the missile without a secret code. Even with the code, multiple procedures must be followed in sequence, including two people turning keys simultaneously in two different parts of the room. The system is designed so that any irregularity makes it impossible to fire the missile at all. Similarly, the body's cells have multiple safety-catch mechanisms. If failure of these mechanisms is detected, other mechanisms stop cell growth. When, despite all previous safeguards, cells grow at an inappropriate rate, still other mechanisms cause the aberrant cells to self-destruct.

A recently discovered gene called p53 is the best example. It makes a protein that protects against cancer by regulating the expression of other genes. In certain circumstances it can shut down cell growth or even make the cell self-destruct. If a person inherits one abnormal copy of the gene that makes this protein, anything that happens to the other copy can lead to catastrophe. The p53 gene is abnormal in fifty-one types of human tumors, including 70 percent of colon cancers, 50 percent of lung cancers, and 40 percent of breast cancers. As John Tooby and Leda Cosmides have pointed out, however, such genetic abnormalities are not necessarily present in the tumor. Cells are often studied after they have lived for years in tissue culture, an environment that may select for genetic abnormalities that increase the rate of cell division.

In addition to these various anticancer mechanisms operating in cells, there are those that operate between them. They detect misbehavior in their neighbors and secrete substances that inhibit the misbehavior. Finally, there is the immune system, which may bring a host of weapons into play against an incipient maladaptive growth as soon as it finds a difference between it and normal tissue. A detectable cancer must somehow have achieved the highly improbable feat of getting past these many layers of defense. Unlike a parasitic worm or infectious bacterium, it cannot draw on a long history of accumulating its own defenses against the host's defenses. It is entirely the product of chance alterations in the cellular regulatory machinery. What cancer has on its side is mainly the astronomical number of chances it gets to achieve success against the immense odds.

CANCER PREVENTION AND TREATMENT

To avoid contracting cancer, the first thing you want to do is to pick your parents wisely. Susceptibility to cancer, like so many other diseases, is hereditary. This is true both in general and for particular forms of cancer, most notably for some rare childhood cancers and those of the breast and colon. Members of families in which such cancers have occurred frequently may have twenty to thirty times the likelihood of contracting them as those in cancer-free families. Even when controlled for family members' tendency to experience similar environmental conditions, the evidence for predisposition for certain kinds of cancer is strong. Mice can be bred to form cancer-prone stocks in which one cancer-control mechanism is already missing in every mouse. This enormously increases the likelihood of one or more kinds of cancer. Some human cancers are inherited in the same way.

Another good way to reduce the likelihood of cancer is to live dangerously: die young, and you are unlikely to get cancer. The fact of senescence means that the environment of any cell and its regulatory capabilities are deteriorating. Hormonal and local regulation of cell growth and proliferation, like all other aspects of adaptive performance, becomes less effective as we go through that terminal life-history stage known as adulthood. The cell itself ages, and as the cardiovascular and digestive and excretory systems deteriorate, it will be ever less well supplied with nutrients and other essentials and ever less effectively unburdened of its waste products. An inevitable consequence is that its potential for growth and cell division is ever less well regulated. Maladaptive growths become steadily more likely to occur and spread unchecked.

The increasing incidence of cancer with age illustrates an important evolutionary principle. Adaptations work best in the circumstances in which they were evolved. Our cancer-control adaptations and other vital functions were not evolved to keep an eighty-year-old alive. The body of anyone that old is an abnormal environment for human genes and their products, one that rarely existed in the Stone Age. More generally, just about all the adverse effects of modern environments, as discussed in Chapter 10, can be expected to increase the incidence of cancer: X rays and other ionizing radiations, novel toxins, unnatural levels of exposure to natural toxins (such as nicotine and alcohol), and abnormal diets or other lifestyle factors.

Injuries and infections anywhere in the body can interfere with cancer-controlling mechanisms not only at the site of the problem but also at distant sites in the body. Bacteria can increase the cancer rates of infected tissues, but viruses are more likely to have such effects. One reason is that a virus is not very different from a single gene in a human cell and can sometimes settle into a niche on a chromosome as if it belonged there. From such a position it can readily subvert the normal machinery of the cell. Viruses, especially HIV, attack the immune system and have the incidental consequence of impairing that system's ability to attack cancer. Like bacteria and larger parasites, viruses can also produce toxins that weaken cellular-control mechanisms.

The connections between environmental causes and certain cancers are sometimes easy to understand. Food that has an abnormally high concentration of salt or alcohol or is loaded with the carcinogens of smoked or broiled meats will contact the stomach cells and increase the risk of stomach cancer. The chemicals in tobacco smoke likewise directly influence lung cells. Sunshine damages the genes in skin cells and leads to melanoma. The mechanism by which a high-fat diet contributes to breast or prostate cancer must depend on more subtle effects than simple contact with the substance in the food. The same can be said for the association between smoking and bladder cancer.

Even after a tumor becomes detectable and produces alarming symptoms, natural control mechanisms, especially immunological factors, will still be at work. They may still win the contest or at least slow the maladaptive growth or prevent its spread to other sites. Even if never cured, some untreated cancers take many years to incapacitate the victim. On rare occasions, apparently incurable cancers just go away.

Many aspects of the contest between a cancer and its victim resemble those between pathogen and host, and the need for such functional categories as cancer-growth adaptations and efforts to suppress them is evident. A cancer is a cellular renegade that has rebelled against the polity of the body and can be regarded as a parasite pursuing its own interests in conflict with the host. Unlike an infectious pathogen, a cancer's success can never be more than short-term, because it has no way to disperse to other hosts, and the host's death means its death too. The same is true of the normal cells from which the cancer arose. When the host dies, the only surviving genes will be those of the host's germ-line cells that have already been passed on to the next generation.

Cancer is a collective term for maladaptive and uncontrolled tissue growths of all kinds. Cancers can arise from any cell types that retain the capacity for growth and division, and cancers of each cell type result from a variety of initiating causes and failures of suppression mechanisms. It is not surprising that cancer has proven difficult for medical science to master, and it is unlikely that one general cure will ever be found. We are making rapid progress, however, and a better understanding of cancer as a contest between renegade cells and the host will surely facilitate more progress.

CANCERS OF FEMALE REPRODUCTIVE ORGANS

Perhaps the best example of a group of related cancers that show the value of a Darwinian approach is cancers of the breast, uterus, and ovary, all of which have recently become much more frequent. Boyd Eaton, a distinguished American researcher in both medicine and anthropology, along with other workers in these fields, has recently brought together a wide range of information to shed light on why these cancers are now so frequent in some human populations and not in others. The evidence is clear that this modern plague is caused in part by the novel reproductive patterns of so many women in the more privileged industrialized societies.

Part of the increase results from the boring fact that cancer is more likely in older people, and more women are now living to old age.

The more interesting finding is that the probability of a cancer of the female reproductive system at any age increases directly in relation to the number of menstrual cycles a woman has experienced. The most likely victim of a cancer of the reproductive tissues is an elderly woman who had an early menarche and late menopause and never had her cycling interrupted by pregnancy and lactation.

From a historical perspective, this is a most abnormal reproductive pattern. Stone Age women, like those of recent hunter-gatherer societies, had quite a different sort of reproductive life history. They had much later maturation and earlier menopause, perhaps in part because they were less well fed and more heavily parasitized than modern women. A Stone Age girl may have experienced menarche at fifteen or later and would probably have been pregnant within a very few years. If the pregnancy miscarried, she would be pregnant again shortly thereafter. If it was successful, it would be followed by a period of lactation of at least two years, possibly four, with associated inhibition of the menstrual cycle. Shortly after weaning (or the death of her infant), she would start cycling and would soon be pregnant again. This would be the pattern until her death or menopause at perhaps about age forty-seven. In this thirty-year period she would have had four or five or maybe six pregnancies and spent more than half of this thirty years lactating. Her total number of menstrual cycles could not have been much more than 150. A modern woman, even if she has two or three children, might easily experience two or three times this number of cycles.

A menstrual cycle is characterized by wide swings of hormone concentrations, and these changes cause cellular responses in the ovarian, uterine, and mammary tissues. These tissue responses are reproductive adaptations, and, like any adaptations, they have costs, in this case increased vulnerability to some forms of cancer. These costs are normally minimized by compensating processes that take place during times when the cycling is interrupted by pregnancy and lactation. If these interruptions never occur, the compensating repairs never occur or are carried out less effectively, and the costs keep accumulating. This is speculation, of course, but it seems very likely that something of the sort must be happening. The undeniable observation is that the more menstrual cycles a woman has, the more likely she is to get a reproductive-system cancer. A more general principle is the adverse effect, for any kind of adaptive mechanism, of conditions other than those in which it evolved. The modern circumstances that lead women to undergo three or four hundred menstrual cycles are no doubt a good example. This evolutionary perspective is not likely to prevent much cancer in women now in their vulnerable years. For them we can do no better than to recommend a general avoidance of environmental hazards such as nicotine and other toxins, both natural and artificial, radiation, and, most important, diets abnormally high in fat.

The long-term implications are more interesting and promising. Obviously it would be both unethical and silly to recommend that girls' growth and maturation should be delayed by an inadequate diet, that they should become pregnant as soon as possible after menarche and frequently again thereafter, and that they should spend a total of perhaps twenty years breast-feeding their babies. Eaton and his coworkers have more enlightened suggestions. What needs to be done is to find out, with carefully conducted research, just how this kind of historically normal life history makes cancer of the reproductive organs less likely. We envision researchers diligently searching for artificial means of achieving the low cancer rates that come naturally to women in hunter-gatherer societies.

We suspect that the artificial means would take the form of hormonal manipulations. Large numbers of women are already using oral contraceptives, which work by artificially affecting tissues in much the same ways as natural hormones do. Different contraceptive medications work in different ways to achieve the desired interference with pregnancy, and they have a diverse array of side effects. With ever more detailed knowledge of the physiological actions of natural and artificial hormones, we should be ever better able to devise artificial ways of mimicking the beneficial effects of Stone Age life histories. This may not be as futuristic and utopian a possibility as it might seem. Eaton and other workers have presented striking evidence that some oral contraceptives can reduce the rates of ovarian and uterine cancer, although not breast cancer. We expect that some sort of hormone treatment will soon be developed to reduce breast cancer as well. None of these comments should be taken to suggest that we should not continue the search for other environmental and genetic causes of cancer. Far from it! We need every bit of knowledge that can help us combat this scourge.

13

SEX AND
REPRODUCTION

Because it is crucial to fitness, you might think that natural selection has smoothly polished the path of sex and reproduction, from the first romantic longings of adolescence to love, marriage, sex, pregnancy, childbirth, and child-rearing. Alas, we all know the truth too well. From unrequited love to lover's spats, premature ejaculation, impotence, lack of orgasm, menstrual problems, the complications of birth, the special vulnerabilities and demands of infants, and the inevitable conflicts between parents, and between parents and their children, reproduction is fraught with strife and suffering. Why does reproduction entail so much conflict and misery? Precisely because it is so crucial to Darwinian fitness. It is at the very core of intense competition and thus causes many problems.

While our main focus in this book is on how evolutionary ideas can help to explain and prevent or cure specific medical diseases, here and in the next chapter we broaden our view somewhat to encompass emotional and behavioral problems that may or may not be considered medical disorders. Some problems associated with reproduction, such as diabetes during pregnancy or sudden infant death syndrome, are clearly diseases, while others, such as jealousy, child abuse, and sexual problems, involve behavior and emotions. However we categorize them, they cause much suffering and make more sense in the light of evolution. Help from Darwinism does not end at the boundary between the medical and the social or educational. Darwinism is relevant to all aspects of human life, not just medicine.

WHY IS THERE SEX?

We begin with a fundamental enigma, one of those wonderful questions that is easy to overlook until you take an evolutionary view of life. Why does sex exist at all? It is costly to fitness in important ways, and many organisms do nicely without it, reproducing either by dividing, like amoebae, or by having females that can lay eggs that develop without fertilization, like aphids. Such creatures have a huge short-term fitness advantage over those who reproduce sexually. Imagine what would happen if a mutation produced a female robin that was perfectly standard in every other respect except that she laid eggs that carried all of her genes but none of her mate's and developed normally without needing to be fertilized. In every generation, all the offspring would be identical females. Compared to a normal female, who can pass only half her genes on to each offspring and who has half male and half female offspring, this mutant strain would increase twice as fast.

So why didn't some parthenogenetic woman, ages ago, flood the world with her progeny and drive us sexual beings to extinction? And why did sex evolve in the first place? Surprising as it may seem, biologists don't yet fully agree about how to answer these questions. Most believe that the function of sex is to introduce variation in offspring, but it remains hard to understand how this variation can be useful enough to outweigh the enormous evolutionary costs of sexual reproduction. Biologists also realize that, in the long run, the recombination of genes during sexual reproduction may prevent an otherwise steady accumulation of deleterious mutations, but this does not answer the question of why asexual reproduction does not continually increase in the short run.

Recently, some scientists have proposed that sexual reproduction is maintained by the selective force of the arms race with pathogens. An individual who is genetically identical to many others is vulnerable to any pathogen that discovers the key to exploiting this bonanza of susceptible individuals. If a clone of ten thousand parthenogenetic women are all vulnerable to influenza, they might all be wiped out by the next epidemic, which would claim only some of their genetically diverse competitors. There is growing support for this hypothesis, including several studies that have found asexual reproduction more frequent in species and in habitats with fewer parasites.

THE ESSENCE OF MALENESS AND FEMALENESS

Imagine a time hundreds of millions of years ago, when cells had begun to exchange genetic material to provide variation but before the development of recognizable eggs and sperm. Such haphazard exchange of genetic material is fraught with conflict. A gene that can get itself donated to many other cells has a major fitness advantage, while one that allows itself to be replaced by genes from other cells is at a substantial disadvantage. The successful gene must get itself into new cells, yet not be displaced by incoming genes. In all organisms above the bacterial level, genes from different individuals are rarely allowed to enter. Genetic recombination is instead accomplished by the production of specialized sex cells (gametes) that can be sent off with half the genes needed for the initiation of a new individual. When two such cells find each other, they unite to produce a new organism with equal genetic contributions from each parent.

Gametes face two difficulties. First, they must have sufficient energy stores both to endure until they merge with another gamete and to nourish a developing embryo. Second, they must find another gamete. Large gametes may have abundant energy stores, but they are expensive to make. Small gametes can be produced in enormous numbers at moderate cost, but they can't survive for long and have nothing to spare for nourishing an embryo. Middle-size gametes sacrifice numbers for larger but still inadequate nutrient supplies and are eliminated by natural selection. Multicellular organisms thus produce only large gametes, which we call eggs, and small ones, which we call sperm.

The next difficulty in understanding human sexuality is why there should be not only two kinds of gametes but two sexes. In other words,

why should there be males that produce sperm and females that produce eggs, rather than hermaphrodites that produce both? Many animals and most plants are hermaphrodites, with both eggs and sperm
produced by the same individual. The consensus among biologists is
that hermaphroditism can be expected when the same adaptations can
serve both sexual functions. Big, bright petals on a flower, for instance,
may attract an insect that both brings pollen that fertilizes the plant's
eggs and picks up pollen to fertilize other plants' eggs. As expected,
most flowering plants are hermaphrodites. In mammals, there is a
dearth of double-duty adaptations. A penis and secondary characteristics such as antlers serve male functions only. A uterus and milk glands
serve only female functions. An individual that invested its limited
resources in both male and female strategies would not be much good
as either. No species of mammal is hermaphroditic.

The investment a female makes in an egg is many times what a
male makes in a sperm. Even when the egg is microscopic, as it is in
humans, it is still thousands of times bigger than a sperm, and two
hundred million sperm cells are released in a single ejaculate to
compete to fertilize a single egg. This initial difference in gamete
expense is perpetuated and magnified. If most of the eggs produced
are fertilized, most of the nutrients put into them will go to the
resulting young. If most of the more numerous sperm die from not
being able to fertilize an egg, nutrients put into them will seldom
benefit an offspring. Extra nutrients in a sperm would be more
likely to retard its swimming and be a handicap in competing for
the limited number of eggs.

If an animal releases eggs into the water, it becomes advantageous
for the female to postpone their release until conditions are ideal and
abundant sperm are nearby. If she can wait to pick a specific male, so
much the better. Genes from a robust, healthy male may give her offspring an advantage. If she can induce males to fight over her or
otherwise display their prowess, she will better her odds of picking
the best possible mate. By retaining the eggs inside her until they are
fertilized, she maximizes control over who fertilizes them, wastes
fewer eggs that are never fertilized, and can protect the eggs to a later
stage of development after fertilization. People automatically think of
internal fertilization as meaning internal to the female, but logically
this need not be. When seahorses copulate, a female lays eggs into a
male's brood pouch, analogous to a mammalian uterus, where the
young develop to an advanced stage. This sort of development inside

the male is exceptional in the animal kingdom. The small size and mobility of sperm cells make it easier for evolution to produce adaptations for getting sperm into a female rather than eggs into a male.

Since the fertilization of a human egg takes place inside the mother, this puts her in charge of the process. It also increases her control over which male will fertilize her eggs. As with females of other species, it is in her reproductive interest to look for males with demonstrable evidence of health and vigor. If females start selecting males with a particular characteristic, such as the huge, colorful feathers of the peacock or the large antlers of an Irish elk, a process of runaway selection may ensue. Males with the characteristic have an advantage simply because females choose them, so females prefer them in order to have sons that the next generation of females will prefer, thus selecting for still more of the characteristic and giving well-endowed males a still greater advantage and a still greater desirability to females. This positive feedback loop elaborates the trait to the point where it may be severely detrimental to the everyday functioning of the males. The poor peacock can hardly fly, and the Irish elk's antlers became so heavy and unwieldy they have been thought responsible for the species' extinction. This is a fine example of how natural selection may create traits that are by no means helpful to the individual or its species, only to the individual's genes. Helena Cronin, in *The Ant and the Peacock*, gives an exquisite history of this idea and of the reluctance of male scientists to acknowledge the power of female choice and its burdensome effects on males.

If there is internal fertilization, the young can presumably be released at the optimal stage. Optimal for whom? Mother? Baby? Father? We'll come to that soon. Exactly how long the young are retained is a life history feature very much subject to natural selection. With the nine-month human pregnancy, in which an offspring grows from a microscopic mite to an infant of several kilograms, a mother's investment in each baby is vastly larger than that of the father. On the other hand, she is sure the baby is hers, while her mate may well be uncertain. This uncertainty means that male expenditures of time and energy caring for the offspring will generally have a more dubious payoff than similar investments by females. The initial tiny difference in the cost of sperm versus the cost of an egg is greatly amplified by human reproductive physiology and leads, as we will see, to different reproductive strategies for males and females.

Girls and boys are born in nearly equal numbers, as we explained in Chapter 2, because individuals of whichever sex is in excess will have lower reproductive success, on average. Selection therefore constantly shapes parents who have offspring of the scarcer sex, thereby equalizing the sex ratio in the long run. From the standpoint of maximizing collective reproduction, this is inefficient. It takes only a few men to keep a large number of women reproducing at whatever rate would maximize the women's reproductive success. This is a clear illustration of the greater importance of lower levels of selection relative to higher (group) levels. If selection at the group level were at all important, the sex ratio would be biased toward females.

This is not a matter of merely academic interest. In India, a cultural preference for males has combined with a proliferation of ultrasound imaging machines, which allow the determination of the sex of a fetus, to severely distort the sex ratio. More than 90 percent of abortions in India are now of female fetuses, and the sex ratio in the general population is beginning to show an imbalance. Similarly, in many areas of China, where population limitation campaigns restrict a couple to one child, that child is a boy more than 60 percent of the time. In the long run such imbalances will be tempered by natural selection, but in the coming generation they will have unpredictable political and social consequences. Our guess is that the excess men will compete vigorously and the scarce women will gain social power with remarkable speed.

CONFLICT AND COOPERATION BETWEEN THE SEXES

Conflict between the sexes is not continuous. Men and women *can* get along, sometimes for whole days at time, even weeks. This harmony is, however, inevitably disrupted by conflicts that originate in the differing reproductive interests and strategies of men and women. From the original difference between the tiny sperm and the larger egg, whole separate worlds of conflicting strategies have emerged to ensnarl our lives. Women can have a limited number of babies, usually four to six, rarely even as many as twenty according to the record books. Men can, however, have hundreds of children and have done so in cultures where a combination of surplus resources and social stratification made it possible for some men to have harems of hundreds of women while many others lacked even a single mate. These exceptional cases are extreme examples of the principle that the number of offspring may vary more widely for men than women. This difference arises from a woman's unavoidably high investment in both time and calories for a single baby, compared to a man's minimal expense of a few minutes and a single ejaculate.

These differences mean that men and women can and do use different kinds of strategies to maximize their Darwinian fitness. A woman can maximize the number of her genes in future generations by finding and keeping a man who will care and provide well for her and her children and who is disinclined to invest in other women. Men can use a similar strategy by finding and keeping a woman who is fertile, inclined to take good care of her children, and disinclined to mate with other men. Men also have another strategy not available to women, that of inseminating many women while providing little or no support for them and their babies. None of this implies that men and women think through their options in order to arrive at conscious strategies to maximize their reproductive success, and it certainly implies nothing about how people ought to act. Nonetheless, natural selection has inevitably shaped our emotional machinery in ways that maximize our reproduction—or that would have in Stone Age circumstances.

MATE PREFERENCES

The problems that result from these divergent strategies are evident in courtship choices. Females of all species do best if they can find a male who offers good genes and abundant resources. Thus, when females can choose, males compete to prove their abilities in contests that range from the familiar butting contests of deer and sheep to the deep braggadocio of the bullfrog. In other species the female mates with the male with the biggest nuptial gift, usually an insect or other source of protein, sometimes the male himself, as when the preying mantis male is eaten by the female even as he copulates with her. The male mantis might try harder to escape, but since he is unlikely to find another mate, he probably maximizes his own reproductive success by donating his bodily protein to the female, who can use it to give more to their offspring.

Men, while notoriously less choosy than women, still have strong preferences. A man maximizes his reproductive success by mating with a woman who has been healthy and successful (indicating good genes) and who is maximally fertile (indicated mainly by being in the peak reproductive years), uncommitted (indicated by lack of prior offspring), and able and motivated for mothering. As University of Michigan psychologist David Buss puts it:

> Imagine a state in which human males had no mate preferences aside from species recognition and instead mated with females randomly. Under these conditions, males who happened to mate with females of ages falling outside the reproductive years would become no one's ancestors. Males who happened to mate with females of peak fertility, in contrast, would enjoy relatively high reproductive success. Over thousands of generations, this selection pressure would, unless constrained, fashion a psychological mechanism that inclined males to mate with females of high fertility over those of low fertility.

So both sexes can increase their fitness by choosing their mates carefully, but they choose different characteristics. Males are relatively more interested in fertility and sexual loyalty, females in good genes and resources. In a study of 10,047 people from diverse cultures and religions in thirty-seven countries, Buss has confirmed these generalizations. Earning capacity was significantly more important to women than to men in all but one of the thirty-seven samples. Youth and appearance were relatively more important to men, and in twenty-three of the thirty-seven samples, men valued chastity significantly more highly than women did, while there was no culture in which the reverse was true.

Mate choice is especially complicated in the human species, where parents mate repeatedly and both provide care for the young. These circumstances mean that a woman faces the risk of being deserted and so must not only assess the current status of her mate but must also try to predict his ability and willingness to stay and provide for her and their offspring. An enduring bond and continuing investment by the man mean that he now also runs a new risk compared to most other primates, that of being cuckolded. He therefore must assess the likelihood that his prospective mate will mate with other men, thus exposing him to the possibility of unwittingly investing in a woman who may be carrying another man's baby and, later, in the offspring of another man.

To succeed, an individual must predict the prospective mate's future behavior, an iffy task at best. Both sexes look for indicators of loyalty and willingness to invest in offspring. Amotz Zahavi, an Israeli biologist, has suggested that these pressures might explain some otherwise mysterious conflicts by a mechanism he has called "testing of the bond." By provoking the prospective partner, he suggests, one can assess his or her willingness to continue to deliver resources and loyalty in the face of future difficulties. Do lovers have spats to test each other? Zahavi provides examples from the world of courting birds to support his theory. Female cardinals, for example, peck and chase wooing males and allow mating only after long persecution. Their subsequent bond lasts for season after season. No one has yet looked in detail at human courtship to see whether we do the same.

Now we return to look at the strongest finding in the Buss study. Despite their differences, both sexes from cultures across the globe consistently agreed on the two most important characteristics they would look for in a mate: (1) kindness and understanding and (2) intelligence. Why do both sexes most of all want a caring and capable partner? For an answer we need to understand why there is such an institution as marriage. Why do men and women in every culture form long-lasting sexual and parenting associations while most other primates have very different kinds of mating systems? This question cannot be answered with certainty, but human patterns of food gathering and child rearing are certainly important parts of the explanation. In the natural environment, one caretaker cannot easily raise a child. Children are, for many years, too helpless and heavy to be taken on long foraging trips. In order to succeed, they need instruction in the ways of their culture and help in negotiating the group hierarchy. In short, each child is so expensive that it may take more than one individual to raise it. To the extent that both parents have all their children in common, they should have minimal conflicts of interest— except, that is, those conflicts that arise from obligations to other relatives. Problems with in-laws are entirely expectable, because helping in-laws directly benefits the genes only of the spouse, not one's own.

DECEPTIVE MATING STRATEGIES

Mating without caring for the offspring benefits men's reproductive interests more than women's. This is consistent with some other aspects of human sexual patterns. First, prostitution is mainly a female profession. While erotic pleasures are possible for both sexes, the balance of supply and demand is such that everywhere men are willing to pay for sex while women rarely have difficulty finding willing sex partners. Second, the strategies that characterize the singles bar scene begin to make sense. In order to get women into bed, men brag about their ability to protect and provide, exaggerating their exploits and flashing their fake Rolex watches as they swear that they are in love forever. Experienced women are rarely completely taken in by this charade, but these patterns of male deception nonetheless seem to work. Men often accuse women of using the converse deceptive strategy, receiving expensive gifts with excited sexual interest and then, later, indignantly expressing surprise that he could have imagined her to be "that kind of woman." For thousands of years, physicians have called this kind of emotional behavior pattern "hysteria." This name arose because commonly associated physical symptoms such as abdominal pain and psychogenic paralysis were thought to result from the wanderings of the womb through the body. Had physicians usually been women, they might never have invented the dubious diagnosis of "hysteria." Instead, women doctors, observing the deceptive mating strategies of men, might have attributed the characteristics of cads to an overly mobile prostate gland and called it "prostateteria."

REPRODUCTIVE ANATOMY
AND PHYSIOLOGY

The human female's reproductive cycles are quite different from those of other primates. Many female primates advertise their fertile periods with odors, bright patches of skin, and changed behavior. These advertisements are useful communications that increase competition and courtship by males during the females' fertile period and discourage sexual harassment at other times. In human females, ovulation is not only unadvertised, it seems to be carefully concealed. The scheduling in women is also different, with human ovulation regularly repeated at about twenty-eight-day intervals, while most primates ovulate only once or twice a year, often in synchrony with the cycles of other females they are associated with. At the end of the cycle, if there is no pregnancy, the human female loses a considerable amount of blood in the menstrual flow. Human sexual activity is not confined to brief fertile periods but occurs throughout the cycle, with substantial time and energy spent on frequent sexual intercourse. Female orgasm in most primates is either absent or brief and inconspicuous, but in humans it is common and may be intense.

While the details remain very much at issue, there is a growing consensus that all these facts fit together. The key is that the woman and her mate both benefit if he is frequently present instead of away for weeks and months at a time. If her cycles were obvious, he could maximize his reproduction by inseminating her only at fertile times, but because he cannot tell when she is fertile, he must stay nearby and copulate at frequent intervals. If early Stone Age women, with their enlarging mental capacities, could know when they were fertile and connect sex with the pain of childbirth, they might avoid their partners at those times and thus decrease their reproductive success. Here is a possibility, first suggested by ornithologist Nancy Burley, where not knowing something may be good for one's fitness. Concealed ovulation also protects the woman somewhat from being impregnated by men more powerful than her mate since such men cannot know when she is fertile and take advantage of her only at that time.

The average frequency of human intercourse, every three days or so, is high enough to make it likely that an ovulation will result in a pregnancy. As we noted before, however, this continuous sexual activity could also mean that bacteria and viruses can hitch regular free rides deep into the woman's reproductive tract. One defense against such infection is the plug of mucus at the cervix that blocks sperm from ascending except during two or three fertile days a month, when the fibrils in the mucus align to make channels just wide enough for the sperm to swim up into the uterus. As suggested by Margie Profet, menstruation may be another defense to kill pathogens and sweep away the beginnings of infection (see Chapter 3). In the natural environment, of course, most women would experience far fewer menstrual cycles, since they would not cycle while pregnant or lactating, which would be most of the time. Anemia from loss of menstrual blood is another of the many problems that result largely from novel aspects of our environment, such as celibacy and effective contraception.

Men are also different from some other male mammals in having testicles permanently lodged in a scrotal sac outside the body proper. This is a precarious location for organs of such vital importance, so there must be a good reason for it. One clue is the infertility that many men experience from wearing tight underwear, which increases the temperature of the testicles. Anatomic examination shows that the veins bringing blood back into the body from the testicles wrap around the artery in a way that provides an effective countercurrent heat exchange mechanism to keep the testes cool. Why sperm cannot be formed at regular body temperature is an unsolved mystery. Men must keep their testicles cool and functioning all the time because fertile women may be available at any time.

The testicles of different primates vary greatly in size, and much of this variation can be explained by differences in mating patterns. A female chimp mates with several males, while female gorillas and orangutans mate with only one male. Because the reproductive success of the male chimpanzee depends not only on inseminating many females but also on the success of his sperm in competing with other sperm to fertilize the egg, natural selection has increased the number of sperm chimp males make as well as the size of their testicles. Gorillas, despite their large size and fearsomeness, have testicles that are about one-fourth the weight of the average chimpanzee testicles. In general, the relative testis weight is high for species in which females often mate with multiple males and low in those with little sperm competition. Where do humans fall? In between but toward the side of less sperm competition. It appears that multiple matings have, however, occurred often enough during human evolution to select for testicles slightly larger than those of species with reliably monogamous mating patterns.

Two British researchers, Robin Baker and Robert Bellis, have taken this topic of sperm competition much further. They note that human sperm in a single ejaculate are of several different kinds, some of which are incapable of fertilizing an egg. Many of these sperm are designed, they argue, specifically to find and destroy any sperm from other men. They have also shown that the volume of ejaculate collected in condoms from monogamous couples increases not merely with the amount of time since the last ejaculation but also with the amount of time the couple have been apart. This suggests an adaptation to increase sperm output when it may be needed to compete with sperm from another man. If confirmed, this will demonstrate that selection has designed our sexual machinery to compete in many different ways and at very close quarters.

JEALOUSY

However understandable jealousy may be, either in the theory of natural selection or in our intuitions, it has surely been responsible for a large part of the world's miseries. Perhaps the ill will and bloodshed caused by Helen's desertion of Agamemnon for Paris, as described by Homer, need not be taken literally, but it is not an implausible account of the emotions such an event could arouse. Canadian psychologists Martin Daly and Margo Wilson have convincingly demonstrated that a large proportion of the murders of women arise out of male jealousy. Othello's lethal frenzy and Desdemona's tragic death have all too many parallels in real life. More commonly, jealousy merely fuels marital battles that stop short of murder but lead to traumatic divorces and all their tragic consequences. In a few individuals the extremity of these feelings and false beliefs that the partner is unfaithful justify the clinical diagnosis of pathological jealousy. To make sense of all this, we must understand the evolutionary origins and functions of the capacity for sexual jealousy.

Maternity is a certainty, but paternity is always a matter of opinion. A man runs the risk of spending years providing for a woman who is having other men's children and of unwittingly caring for children not his own, while women always know who their children are. A man incapable of jealousy would have a greater risk for being cuckolded, with a resulting decrease in reproductive success. Men who threaten potential interlopers and try to prevent their wives from mating with other men have an evolutionary advantage. Genes that predispose to male sexual jealousy will thus be maintained in the gene pool.

While women do not face the same risk, they face others. A husband's wandering affections can lead to a drain of resources and time, to potential loss of the husband, and to the risk of sexually transmitted diseases. Cross-cultural data show enormous diversity in sexual mores, from cultures where extramarital liaisons are tolerated to those where any infidelity is punished by death. However, sexual jealousy is consistently reported to be more intense for men than women.

Sexual jealousy is such a strong influence on human life that it is institutionalized and regulated by custom or formal law in almost all societies. In technologically advanced Western countries men often treat women as property and try to control their sexuality, but in many traditional societies the control may be even more blatant and institutionalized. In some Mediterranean societies, women must demonstrate their virginity with blood on the marital sheet and then are cloistered so they can associate with no men but their husband. In some Muslim societies women must wear robes and veils that make them unrecognizable by men outside the home. In China, women's feet were bound from early childhood to discourage straying. In many areas of Africa it remains routine for girls at puberty to have the clitoris excised and the labia sewn shut. Everywhere men create social institutions to control female sexuality.

What would be the attitude, in our own society, toward a woman who is faithful to her husband 90 percent of the time, but who has another lover for the remaining 10 percent of her sex life? Her husband would have a 90 percent probability of being the father of her next child, and so, from a strictly evolutionary perspective, we would expect him to be 90 percent as good a father to that child as he would if his wife had been perfectly monogamous. Yet in many cultures a single instance of adultery by a woman may be legally considered a justification for total cancellation of the marriage and abandonment of any ensuing child by the woman's husband. Many people seem to think that culture opposes such biological tendencies, but with jealousy, culture and the legal system exaggerate a biological tendency. People who think that laws should oppose our more destructive biological tendencies would presumably want to change the social system in ways that would discourage divorces based on infidelity. What do you think the world will be like if someone invents a pill that cures jealousy?

SEXUAL DISORDERS

People are, to put it mildly, very interested in the quality of their sex lives. This is ultimately because genes that result in behaviors that increase reproduction have been selected for, while genes that make people uninterested in sex have been eliminated. But from this point on, sex becomes more problematic.

The ubiquity of sexual problems is confirmed by a visit to any bookstore. The very existence of rows of sex therapy books documents the unfortunate truth. Sex is a problem not just for a few people some of the time, but for many people much of the time. The books contain strong hints that these problems are not genetic defects, not results of an abnormal environment, but direct products of evolution. Each book has a chapter about premature orgasm in men and another on delayed or absent orgasm in women. There are no chapters about too-rapid orgasms in women or too-slow orgasms in men and no explanations for why men and women differ in this regard. There are chapters on men with fetishes but no mention of similar problems in women, and again, no comment on why the sexes differ in this susceptibility. Some difficulties the sexes share: both are troubled, on occasion, by lack of sexual desire and difficulty getting aroused. And both sexes (but especially men) are troubled by boredom with the same sexual partner. Here, at the heart of reproduction, we find a biological system that seems haphazard at best. Why should men and women have so many and such different complaints?

At the very least, we might expect the evolved regulatory mechanisms to coordinate the orgasms of men and women. But orgasms are not only uncoordinated, they are systematically sooner for men than women. This bias is one of the more unfortunate illustrations of the principle that natural selection shapes us to maximize reproduction, not satisfaction. Imagine the reproductive success of a man who tends to come to orgasm very slowly. He might please his partner, but if the sex act is interrupted or his partner has been satisfied and does not want to continue, his sperm will sometimes not get to where they will do his genes any good. The same forces shape the timing of the female sexual response. A woman who rapidly has a single orgasm may, on occasion, stop intercourse before her partner ejaculates and thus will have fewer offspring than the woman with a more leisurely sexual response.

A closer look reveals that there may be a system to adjust male sexual timing according to the particular circumstances. Premature ejaculation is common mainly in young men, especially when they are in anxiety-provoking situations. According to anthropologists who study hunter-gatherer cultures, the liaisons of young men are often illicit and would be dangerous if discovered by older men. In such circumstances, brevity of the sexual act may be especially adaptive. These ideas are mere speculation now, but they deserve consideration.

PREGNANCY

Pregnancy would seem to be the ultimate in shared goals—a refuge from conflict, perfect unity of purpose between mother and fetus. And the relationship between mother and fetus is about as intimate and mutual as any relationship can be. Nonetheless, because mother and fetus share only half their genes, there is conflict aplenty. Whatever benefits go to the fetus help all its genes. The fetus maximizes its fitness by appropriating whatever maternal resources it can use short of jeopardizing the mother's ability to care for it in the future and her ability to raise full or half brothers and sisters (all discounted by the one half or three quarters of genes they do not have in common).

From the mother's point of view, benefits given to the fetus help only half of her genes, so that her optimum donation to the fetus is lower than the amount that is optimal for the fetus. She is also vulnerable to injury or death from the birth of too large a baby. The fitness interests of the fetus and the mother are therefore not identical, and we can predict that the fetus will have mechanisms to manipulate the mother to provide more nutrition and that the mother will have mechanisms to resist this manipulation.

People sometimes argue that there could be no net advantage to a gene that benefits an offspring at a cost to its mother, because its early advantage would be exactly reversed by the later cost. This is not the way things work out. Suppose, in a population in which maternal and fetal interests are served equitably, a gene arises that increases fetal nutrition slightly, at a slight cost to the mother. A fetus that enjoys that advantage can avoid the cost half the time when it grows up, because only half its offspring will carry the gene. Also, even more obviously, it will pay the cost only if it is female. So the cost would be paid in only about 25 percent of the pregnancies of the next generation. There are additional complexities—which we will not go into—but such quantitative considerations led Harvard biologist David Haig to expect conflict between parent and offspring, even though the ideal contribution from the mother's perspective may be only slightly less than the ideal for the fetus.

Unfortunately, these slight differences create major conflicts. The fetus may be striving mightily to glean an extra few percent of nutrient delivery from the mother, while the mother tries just as hard to prevent this. When the balance of power is disrupted because one participant's efforts are seriously impaired, medical problems arise. For example, the fetus secretes a substance, *human placental lactogen* (hPL), that ties up maternal insulin so that blood glucose levels rise and provide more glucose to the fetus. The mother counters this fetal manipulation by secreting more insulin, and this makes the fetus secrete even more hPL. This hormone is normally present in all human bodies, but in a pregnant woman it can reach a thousand times the normal concentration. As Haig points out, these raised hormone levels, like raised voices, are a sign of conflict.

If the mother happens to be deficient in her production of insulin, this can cause gestational diabetes, possibly fatal to the mother, and therefore to the glucose-greedy fetus itself. The fetus would have been well advised to curtail its secretion of hPL, but all it can do is play the odds. The average mother is thoroughly competent to produce enough insulin to avoid diabetes, even when flooded by fetal hPL.

The evolutionary theory of parent-offspring conflict was worked out many years ago by Robert Trivers, but it was only in 1993 that David Haig applied it to the workings of human pregnancy. It is also only recently that an unexpected but highly relevant genetic phenomenon came to light. Experiments, mainly with mice, have shown that the genes need not rely on the lottery of sexual reproduction to avoid the later costs of special benefits in fetal development. They may resort to *genetic imprinting*, whereby a gene is somehow conditioned by its parent either to start acting immediately or to avoid acting in the offspring. Genes from a father may be imprinted so they side with a fetus in the conflict with the mother. These same genes, when they come from a mother, may be imprinted so they have no such effect. The relevance of this to human pregnancy remains to be determined, but in mice, genes imprinted by males produce a fetal growth factor and other genes imprinted by females produce a mechanism for destroying that growth factor. Such evidence suggests that it may not be farfetched to view the womb itself as the battleground on which genes play out their interests at the expense of our health.

Aside from diabetes, another scourge of pregnancy is high blood pressure. This is called preeclampsia when it gets severe enough to damage the kidneys so that protein is lost in the urine. Haig has suggested that this too may result from conflict between the fetus and the mother. In the early stages of pregnancy, the placental cells destroy the uterine nerves and arteriolar muscles that adjust blood flow, and this makes the mother unable to reduce the flow of blood to the placenta. If something constricts other arteries in the mother, her blood pressure will go up and more blood will therefore go to the placenta. The placenta makes several substances that can constrict arteries throughout the mother's body. When the fetus perceives that it is receiving inadequate nutrition, the placenta releases these substances into the mother's circulation. They can damage the mother's tissues, but selection may have shaped a fetal mechanism that takes this risk in order to benefit itself even at the expense of the mother's health. Data on thousands of pregnancies show that moderate increases in maternal blood pressure are associated with *lower* fetal mortality, and that women with preexisting high blood pressure have larger babies. Further support is provided by findings that preeclampsia is especially common when the blood supply to the fetus is restricted, and that the mother's high blood pressure results from increased resistance in the arteries, not from increased pumping by the heart.

We wonder if the same mechanism may explain some adult high blood pressure. Low-birth-weight infants are especially likely to develop this condition as adults. If genes that are expressed in the fetus to make substances that increase the mother's blood pressure continue to be active, this could cause high blood pressure later in life.

From a traditional medical perspective, these explanations for diabetes and high blood pressure in pregnancy are revolutionary, and unproven, but we suspect they may well prove correct. If so, they provide extraordinary evidence for the power of looking at life from the gene's point of view, for the ubiquity of biological conflicts of interest, and for the practical utility of an adaptationist approach to disease.

Human chorionic gonadotropin (hCG) is another hormone made by the fetus and secreted into the mother's bloodstream. It binds to the mother's luteinizing hormone receptors and stimulates the continued release of progesterone from the mother's ovaries. This hormone blocks menstruation and lets the fetus stay implanted. hCG seems to have originated in the contest between the fetus and the mother over whether the pregnancy should continue or not. Up to 78 percent of all fertilized eggs are never implanted or are aborted very early in pregnancy. The majority of these aborted embryos have chromosomal abnormalities. Mothers seem to have a mechanism that detects abnormal embryos and aborts them. This adaptation prevents continued investment in a baby that would die young or be unable to compete successfully in adult life. It is advantageous for the mother to cut her losses as early as possible and start over, even if this means culling a few normal embryos in order to avoid the risk of nurturing an abnormal one. The fetus, by contrast, does everything it can to implant itself and to stay implanted. Producing hCG is an important early strategy for the fetus to further this goal.

It seems likely that high hCG levels are somehow detected and interpreted by mothers' bodies as a sign of a viable fetus—if it can make enough hCG, it is probably normal. So the embryo, to demonstrate its fitness to the mother, must now make greater amounts of hCG, levels that say as loud as they can, "I am the makings of a great baby." It is also conceivable, as Haig points out, that these high levels of hCG are a cause of nausea and vomiting in pregnancy. Do you think this an alternative to Profet's morning-sickness theory, summarized in Chapter 6? Not if you understand the distinction between proximate and ultimate causes (Chapter 2). The hCG effect could be part of the adaptive machinery that deters ingestion of toxins. Conversely, it may just be an incidental consequence of high hCG levels. Only a well-designed investigation can resolve this issue.

BIRTH

The large brains and small pelvic openings of humans have combined to make birth especially stressful and risky. As we noted in Chapter 9, it would be far better if the baby could be born through an opening in the abdominal wall, as occurs artificially in a cesarean section, but historical constraints make that impossible, and the baby must still squeeze through the pelvis. The relative immaturity and helplessness of human babies compared to those of other primates are an unavoidable cost of being small enough to be born, but the dangers nonetheless remain for both baby and mother.

Wenda Trevathan, an anthropologist at New Mexico State University, notes that while other primates go off alone to give birth, human mothers often seek companionship and support. She suggests that this may in part be explained by the unusual birth orientation of human babies. In contrast to those of other primates, human babies normally emerge facing backward, so that if the mother were to try to finish a difficult labor by pulling on the baby, she might injure it. The presence of a helper at birth greatly decreases the risk. Even in modern times, the simple presence of a supportive woman during birth can reduce the rate of cesarean section by 66 percent and the use of forceps by 82 percent. Six weeks after birth, mothers who had a helper at birth are less anxious and breast-feed more easily than mothers who gave birth without a helper.

After the baby is born, a modern obstetrician or midwife helps extract the placenta and tries to minimize bleeding. Oxytocin is a natural hormone stimulated by nursing that constricts uterine blood vessels at birth, and injections of extra oxytocin have stopped excessive bleeding and saved thousands of lives. Doctors cannot always predict who will bleed excessively, and oxytocin administration is now part of the delivery routine. There has, however, been little research on the possibility that such routine administration of oxytocin might disrupt other mechanisms.

In some species, notably sheep, birth by cesarean section usually results in the mother not accepting the offspring as her own. A ewe will kick and butt her lamb born by cesarean section. During normal birth, pressure on her vaginal walls stimulates the release of oxytocin, which activates a brain mechanism that makes the mother bond to the first lamb she sees in the next few minutes. Administration of a dose of oxytocin enables a ewe to bond normally to a lamb delivered by cesarean section. We don't know whether oxytocin plays a similar role in human bonding. Because human mothers seem to attach normally to cesarean babies born by cesarean section, it seems that oxytocin may not be necessary to bonding by human mothers. Need this mean it doesn't help? Because the issue is so important, and because of the frequency of cesarean sections and the routine administration of large doses of extra oxytocin, further study of the positive and negative effects of this hormone is needed.

INFANCY

When the baby first nurses at the mother's breasts, they secrete not milk but colostrum, a watery liquid full of substances that protect the baby from infection. In a few days, the real milk comes in, which also contains a variety of substances that protect the baby far better than anything in infant formula. Much has been said about the benefits of natural breast-feeding, and we will not belabor the point, except to note parenthetically how completely nonadaptive human behavior can be in the modern environment. For instance, four of Mozart's six children died in the first three years of life—tragic but not surprising when we learn that they were fed mainly sugar water.

Many babies now spend a few extra days in the hospital because they are jaundiced. The yellow color results from high levels of bilirubin, a by-product of the breakdown of hemoglobin. At the time of birth, fetal hemoglobin, which is well suited to the intrauterine environment, is being replaced by the adult form, which is better suited to life outside the womb. If the liver gets behind in processing the great onslaught of hemoglobin derivatives, a certain amount of jaundice is both understandable and unremarkable.

Physicians first recognized the dangers of high levels of bilirubin in those babies whose blood cells had an Rh antigen that is attacked by their mother's antibodies. The rapid breakdown of blood cells and resulting high bilirubin levels sometimes caused permanent brain damage. Today this can usually be prevented by administering substances that prevent the mother from developing Rh antibodies or by giving the baby an exchange transfusion at birth. But many babies who do not have Rh antigens also have visible jaundice at birth. To prevent any possibility of brain damage, such babies are often treated with exposure to bright light, which changes the bilirubin in the skin to a form that can be excreted in the urine, thus hastening the disappearance of jaundice.

So far it looks as if the high bilirubin levels at birth are simply a glitch in the mechanism, one we can fortunately circumvent by routine medical treatment. John Brett at the University of California at San Francisco and Susan Niermeyer at the Children's Hospital in Denver have taken a more careful evolutionary look at this situation. They note that the first breakdown product of hemoglobin is biliverdin, a water-soluble chemical that is excreted directly in birds, amphibians, and reptiles. In mammals, however, biliverdin is converted to bilirubin, which is then transported throughout the body bound to the blood protein albumin. Furthermore, bilirubin levels at birth are under partial genetic control and therefore could be lowered by natural selection if that were beneficial. This led Brett and Niermeyer to suspect that high bilirubin levels at birth might be adaptive. As they put it, "Given that all babies will be jaundiced well above the adult level within the first postnatal week and over half will be visibly jaundiced, it seems difficult to imagine that something is wrong with all of these infants." Further investigation revealed that bilirubin is an effective scavenger of the free radicals that damage tissues by oxidation. At birth, when the baby must suddenly start breathing, the arterial oxygen concentration becomes three times as great, with concordant increases in damage from free radicals. Adult levels of defenses against free radicals are only gradually implemented during the first weeks of life, as the bilirubin levels decrease. If Brett and Niermeyer are correct, we need to rethink our treatment of jaundice of the newborn, perhaps saving millions of dollars in unnecessary treatment each year.

The risks of light treatment have been inadequately investigated, but we know that color vision impairments can result from continuous bright light in the first few days after birth. We want to make it clear that the adaptive interpretation of Brett and Niermeyer has not been widely accepted and strongly caution parents against refusing to let their babies have light treatment if their doctors deem it necessary. It would be worthwhile, however, for parents to ask questions and to get second opinions, and for scientists to initiate studies to provide the decisive answers.

CRYING AND COLIC

The baby is home now, and the wonderful joy is punctuated, regularly, day and night, by hours of wails that cannot be ignored. It is easy enough to understand how crying benefits the baby. If it is hungry, thirsty, hot, cold, frightened, or in pain, the baby cries and a parent comes to meet its needs. A baby unable to cry might be seriously neglected. How does the baby's cry affect parents? It gets on their nerves, to put it mildly. Parents do whatever needs to be done to stop the crying, at any time of day or night. Genes that make the cry aversive to parents are selected for because those same genes are in the child, who benefits from the parent's discomfort and resulting aid. The parent suffers, but its genes in the baby benefit—a fine example of the actions of kin selection.

If the baby cries for a good reason, all to the good. But is all crying a call for necessary help? Often it is impossible to find any cause at all, and yet nothing seems to stop the baby's crying. This is the most common reason new mothers consult their pediatricians, who usually call the problem "colic" despite little evidence that gastrointestinal difficulties are responsible. Ronald Barr, a pediatrician at McGill University, has made an intensive study of infant crying. He finds that babies with supposed colic do not cry more often or at special times, just longer each time. This has led him to suggest that such crying is normal, although perhaps prolonged by modern practices such as long intervals between feedings. !Kung women in Africa carry their infants constantly and feed them whenever they cry, at least once and often three or four times per hour for two minutes at each feeding. By contrast, American mothers feed their two-month-old infants approximately seven times a day with an average of three hours between feedings. In an experimental study, Barr asked a group of mothers to carry their babies at least three hours per day. Mothers in that group reported that their babies cried only half as long as those whose mothers did not receive the special instructions.

Barr suggests that frequent crying increases fitness by promoting bonding with the mother and by encouraging frequent feeding, which maintains lactation and prevents any competing pregnancy. This last argument again illustrates the conflict of interests between the parent and the offspring. The frequency of babies "spitting up" may be another instance in which the baby manipulates the mother, in this case to make more milk than is in her interests. Or "spitting up" may be explained as a result of unnaturally infrequent but larger feedings. An examination of the phenomenon in hunter-gatherer societies could provide an answer, but it is not the kind of thing that anthropologists routinely report.

SUDDEN INFANT DEATH SYNDROME (SIDS)

Many a parent's greatest fear is of going to wake the baby and finding it dead in the crib. Sudden infant death syndrome (SIDS) kills more babies than any other cause of death except accidents—1.5 per 1000 babies, or more than 5000 per year in the United States alone. The cause, however, remains unknown. James McKenna, an anthropolo-

gist from Pomona College, has investigated SIDS from an evolutionary and cross-cultural perspective and found that crib deaths are
many times more frequent in modern societies than in tribal cultures.
The SIDS rate is especially high, as much as ten times higher, in those
cultures in which babies sleep apart from their parents instead of in
the same bed. In a series of experiments that simultaneously measured the movements and brain waves of sleeping mothers and their
babies, he found substantial relationships between the sleep cycles of
mothers and babies who sleep together. He suggests that this coordination leads to intermittent arousals that sustain SIDS-vulnerable
babies through periods when their breathing might otherwise cease.
The more fundamental problem, cessation of breathing, may be
related to the extreme immaturity of the human infant's nervous system, the price of avoiding the danger of the birth of babies with too
large a skull to fit through the pelvis. None of this is to say that SIDS
is in any way normal, only that the tendencies that make some infants
vulnerable to it may have been far less dangerous in a natural environment, where mothers usually sleep with their newborns.

WEANING AND BEYOND

Eventually the mother begins to discourage the baby from
nursing. In industrial societies, this usually occurs some-
time in the first year, while in hunter-gatherer cultures nursing lasts an average of three to four years. The interval
between births is critical to maximizing reproduction. If it is too
short, the first infant may still need so much milk and effort that the
next infant will not survive. If the mother waits too long, she is wasting her reproductive potential. As you might expect from our discussions of parent-offspring conflict, this is yet another instance in
which the interests of the mother and the infant diverge. There will
come a time, usually when an infant is two to four years old, when it
is in the mother's genetic interests to conceive again but in the baby's
interests to keep nursing and prevent her from having another baby.
This is the weaning conflict, discussed by biologist Robert Trivers in
his classic paper that first outlined the divergent interests of parents
and their offspring. He noted that weaning conflicts have a natural
end point. Eventually, the baby can do well enough with solid foods

and less aid from the mother that it too will benefit more from having a baby brother or sister (who shares half its genes) than from continuing to monopolize its mother.

During the period of weaning conflict, how can the infant manipulate its mother to continue nursing? Here again Trivers had a brilliant insight. The infant, unable to force the mother to keep nursing, can only use deception, and the best deception is to convince the mother that it is in her best interest to let nursing continue. How can the baby accomplish such deception? Simply by acting younger and more helpless than it really is. Psychologists have long recognized this pattern and named it *regression*, but we believe Trivers has offered the first evolutionary explanation, with implications that are just beginning to be explored.

Parent-offspring conflicts don't end with weaning; they just change their form. For a long period in childhood, conflicts are relatively routine and mild, but come adolescence, all hell breaks loose. Teenagers may want to do everything their own way and insist that no help of any sort is needed. Then, at the least difficulty, they are back into the regression act, apparently helpless and needy and asking for more than the parents want to give. This isn't so surprising, really. It is just the last major episode of parent-offspring conflict in the long drama of development. In a few years the adolescent really will be independent and beginning to look longingly at a potential partner with whom to raise a family and start a new episode in that ongoing drama of adaptively modulated conflict and cooperation called sexual reproduction.

14

ARE MENTAL DISORDERS DISEASES?

I sometimes hold it half a sin
　　To put in words the grief I feel:
　　For words, like Nature, half reveal
And half conceal the Soul within.

But, for the unquiet heart and brain,
　　A use in measured language lies;
　　The sad mechanic exercise,
Like dull narcotics, numbing pain.

　　　　　　　　—Alfred, Lord Tennyson,
　　　　　　　　　In Memoriam, canto V

A young woman recently came to the Anxiety Disorders Clinic at The University of Michigan, complaining of attacks of overwhelming fear that had come out of the blue several times each week for the past ten months. During these attacks, she experienced a sudden onset of rapid pounding heartbeats, shortness of breath, a feeling that she might faint, trembling, and an overwhelming sense of doom, as if she were about to die. A few years ago, such people usually insisted that they had heart disease, but this person, like so many now, had read about her symptoms and knew that they were typical of panic disorder. In the course of the evaluation it came out that she had experienced her first panic attacks at about the same time as she had begun an extramarital affair. When the doctor asked if there might be a connection, she said, "I don't see what that has to do with it. Everything I read says that panic disorder is a disease caused by genes and abnormal brain chemicals. I just want the medicine that will normalize my brain chemicals and stop these panic attacks, that's all."

How times change! Twenty years ago, people who insisted that their anxiety was "physical" were often told that they were denying the truth in order to avoid painful unconscious memories. Now many psychiatrists would readily agree that depression or anxiety can be a symptom of a biological disease caused by brain abnormalities that need drug treatment. Some people, like the woman described above, so embrace this view that they are offended if the psychiatrist insists on attending to their emotional life. The opening lines of an influential review article summarize these changes:

> The field of psychiatry has undergone a profound transformation in recent years. The focus of research has shifted from the mind to the brain . . . at the same time the profession has shifted from a model of psychiatric disorders based on maladaptive psychological processes to one based on medical diseases.

Strong forces have pushed the field of psychiatry to adopt this "medical model" for psychiatric disorders. The change began in the 1950s and 1960s with discoveries of effective drug treatments for depression, anxiety, and the symptoms of schizophrenia. These discoveries spurred the government and pharmaceutical companies to fund research on the genetic and physiological correlates of psychiatric disorders. In order to define these disorders so research findings from different studies could be compared, a new approach to psychiatric diagnosis was created, one that emphasizes sharp boundaries around clusters of current symptoms instead of continuous gradations of emotions caused by psychological factors, past events, and life situations. Academic psychiatrists focus increasingly on the neurophysiological causes of mental disorders. Their views are transmitted to residents in training programs and to practitioners via postgraduate medical seminars. Finally, with the rise of insurance funding for medical care during recent decades and the possibility of federal funding for universal medical coverage in the United States, organizations of psychiatrists have become insistent that the disorders they treat are medical diseases like all others and therefore deserve equal insurance coverage.

Are panic disorder, depression, and schizophrenia medical diseases just like pneumonia, leukemia, and congestive heart failure? In our opinion, mental disorders are indeed medical disorders, but *not* because they are all distinct diseases that have identifiable physical causes or because they are necessarily best treated with drugs. Instead, mental disorders can be recognized as medical disorders when they are viewed in an evolutionary framework. As is the case for the rest of medicine, many psychiatric symptoms turn out not to be diseases themselves but defenses akin to fever and cough. Furthermore, many of the genes that predispose to mental disorders are likely to have fitness benefits, many of the environmental factors that cause mental disorders are likely to be novel aspects of modern life, and many of the more unfortunate aspects of human psychology are not flaws but design compromises.

EMOTIONS

U npleasant emotions can be thought of as defenses akin to pain or vomiting. Just as the capacity for physical pain has evolved to protect us from immediate and future tissue damage, the capacity for anxiety has evolved to protect us against future dangers and other kinds of threats. Just as the capacity for experiencing fatigue has evolved to protect us from overexertion, the capacity for sadness may have evolved to prevent additional losses. Maladaptive extremes of anxiety, sadness, and other emotions make more sense when we understand their evolutionary origins and normal, adaptive functions. We also need proximate explanations of both the psychological and brain mechanisms that regulate and express these emotions. If we find what look like abnormalities in the brains of people who are anxious or sad, we cannot conclude that these brain changes cause the disorder in any but the most simplistic sense. Brain changes associated with anxiety or sadness may merely reflect the normal operation of normal mechanisms.

Knowledge about the normal functions of the emotions would provide, for psychiatry, something like what physiology provides for the rest of medicine. Most mental disorders are emotional disorders, so you might think that psychiatrists are well versed in the relevant scientific research, but no psychiatric training program systematically teaches the psychology of the emotions. This is not as unfortunate as it seems, since research on the emotions has been as fragmented and confused as psychiatry itself. In the midst of ongoing technical debates, however, many emotions researchers are reaching consensus on a crucial point: *our emotions are adaptations shaped by natural selection*. This principle holds substantial promise for psychiatry. If our emotions are subunits of the mind, they can be understood, just like any other biological trait, in terms of their functions. Doctors of internal medicine base their work on understanding the functions of cough and vomiting and the liver and the kidneys. An understanding of the evolutionary origins and functions of the emotions would begin to provide something similar for psychiatrists.

Many scientists have studied the functions of the emotions. Some have emphasized communication, especially University of California psychologist Paul Ekman, whose studies of the human face demonstrate the cross-cultural universality of emotions. Others emphasize the utility of emotions for motivation or other internal regulation, but emotions have not been shaped to perform one or even several functions. Instead, each emotion is a specialized state that simultaneously adjusts cognition, physiology, subjective experience, and behavior, so that the organism can respond effectively in a particular kind of situation. In this sense, an emotion is like a computer program that adjusts many aspects of the machine to cope efficiently with the challenges that arise in a particular kind of situation. Emotions are, in the felicitous phrase coined by University of California psychologists Leda Cosmides and John Tooby, "Darwinian algorithms of the mind."

Emotional capacities are shaped by situations that occurred repeatedly in the course of evolution and that were important to fitness. Attacks by predators, threats of exclusion from the group, and opportunities for mating were frequent and important enough to have shaped special patterns of preparedness, such as panic, social fear, and sexual arousal. Situations that are best avoided shape aversive emotions, while situations that involve opportunity shape positive emotions. Our ancestors seem to have faced many more kinds of threats than opportunities, as reflected by the fact that twice as many words describe negative as positive emotions. This perspective gives the boot to the modern idea that "normal" life is free of pain. Emotional pain is not only unavoidable, it is normal and can be useful. As E. O. Wilson put it,

> Love joins hate; aggression, fear; expansiveness, withdrawal; and so on; in blends designed not to promote the happiness and survival of the individual, but to favor the maximum transmission of the controlling genes.

But much emotional pain is not useful. Some useless anxiety and depression arise from normal brain mechanisms, others from brain abnormalities. Major genetic factors contribute to the causation of anxiety disorders, depression, and schizophrenia. In the next decade, specific genes will no doubt be found responsible for certain kinds of mental disorders. Physiological correlates have been found for all of these disorders, and neuroscientists are hard at work unraveling the responsible proximate mechanisms. The resulting knowledge has already improved the utility of drug treatments and offers the possibility of prevention. This is a bright time for psychiatry and for people with mental disorders. The advances in pharmacologic treatment have come so fast that many people remain unaware of their safety and effectiveness. Treatment is now more effective than the wildest hopes of psychiatrists who went into practice just thirty years ago.

Much confusion attends these advances. The human mind tends to oversimplify this issue by attributing most bad feelings either to genes and hormones or to psychological and social events. The messy truth is that most mental problems result from complex interactions of genetic predispositions, early life events, drugs and other physical effects on the brain, current relationships, life situations, cognitive habits, and psychodynamics. Paradoxically, it now is much easier to treat many mental disorders than it is to understand them.

Just as there are several components of the immune system, each of which protects us against particular kinds of invasions, there are subtypes of emotion that protect us against a variety of particular kinds of threats. Just as arousal of the immune system usually occurs for a good reason, not because of an abnormality in its regulation mechanism, we can expect that most incidents of anxiety and sadness are precipitated by some cause, even if we cannot identify it. On the other hand, the regulation of the immune system can be abnormal. The immune system can be too active and attack tissues it shouldn't, causing autoimmune disorders such as rheumatoid arthritis. Comparable abnormalities in the anxiety system cause anxiety disorders. The immune system can also fail to act when it should, causing deficiencies in immune function. Might there be anxiety disorders that result from too little anxiety?

ANXIETY

E veryone must realize that anxiety can be useful. We know what happens to the berry picker who does not flee a grizzly bear, the fisherman who sails off alone into a winter storm, or the student who does not shift into high gear as a term-paper deadline approaches. In the face of threat, anxiety alters our thinking, behavior, and physiology in advantageous ways. If the threat is immediate, say from the imminent charge of a bull elephant, a person who flees will be more likely to escape injury than one who goes on chatting nonchalantly. During flight, our survivor experiences a rapid heartbeat, deep breathing, sweating, and an increase in blood glucose and epinephrine levels. Physiologist Walter Cannon accurately described the functions of these components of the "fight or flight" reaction back in 1929. It is curious that his adaptationist perspective has never been extended to other kinds of anxiety.

While anxiety can be useful, it usually seems excessive and unnecessary. We worry that it will rain at the wedding next June, we lose our concentration during exams, we refuse to fly on airplanes, and we tremble and stumble over our words when speaking in front of a group. Fifteen percent of the U.S. population has had a clinical anxiety disorder; many of the rest of us are just nervous. How can we explain the apparent excess of anxiety? In order to determine when it is useful and when it is not, we need to ask how the mechanisms that regulate anxiety were shaped by the forces of natural selection.

Because anxiety can be useful, it might seem optimal to adjust the mechanism so that we are always anxious. This would be distressing, but natural selection cares only about our fitness, not our comfort. The reason we are sometimes calm is not because discomfort is maladaptive but because anxiety uses extra calories, makes us less fit for many everyday activities, and damages tissues. Why does stress damage tissues? Imagine a host of bodily responses that offer protection against danger. Those that are "inexpensive" and safe can be expressed continually, but those that are "expensive" or dangerous cannot. Instead, they are bundled into an emergency kit that is opened only when the benefits of using the tools are likely to exceed the costs. Some components are kept sealed in the emergency kit precisely because they cause bodily damage. Thus, the damage associated with chronic stress should be no cause for surprise and certainly no basis for criticizing the design of the organism. In fact, recent work has suggested that the "stress hormone" cortisol may not defend against outside dangers at all but instead may mainly protect the body from the effects of other parts of the stress response.

If anxiety can be costly and dangerous, why isn't the regulatory mechanism adjusted so that it is expressed only when danger is actually present? Unfortunately, in many situations it is not clear whether or not anxiety is needed. The smoke-detector principle, described previously, applies here as well. The cost of getting killed even once is enormously higher than the cost of responding to a hundred false alarms. This was demonstrated by an experiment in which guppies were separated into timid, ordinary, and bold groups on the basis of their reactions when confronted by a smallmouth bass: hiding, swimming away, or eyeing the intruder. Each group of guppies was then left in a tank with a bass. After sixty hours, 40 percent of the timid guppies and 15 percent of the ordinary guppies were still there, but none of the bold guppies had survived.

The psychiatrist's attempt to understand how natural selection has shaped the mechanism that regulates anxiety is conceptually the same as the electronics engineer's problem of determining if a signal on a noisy telephone line is actually information or just static. Signal detection theory provides a way to analyze such situations. With an electronic signal, the decision about whether to call a given click a signal or noise depends on four things: (1) the loudness of the signal, (2) the ratio of signals to noise, (3) the cost of mistakenly thinking that a noise is actually a signal (false alarm), and (4) the cost of mistakenly thinking that a signal is actually a noise (false negative response).

Imagine that you are alone in the jungle and you hear a branch break behind a bush. It could be a tiger, or it could be a monkey. You could flee, or you could stay where you are. To determine the best course of action, you need to know: (1) the relative likelihood that a sound of this magnitude would come from a tiger (as opposed to a monkey), (2) the relative frequency of tigers and monkeys in this location, (3) the cost of fleeing (the cost of a false alarm), and (4) the cost of not fleeing if it really is a tiger (the cost of a false negative response). What if you hear the sound of a medium-sized stick breaking behind that bush? The individual whose anxiety level is adjusted by an intuitive, quick, and accurate signal detection analysis will have a survival advantage.

The analogy with the immune disorders suggests that there might be a whole category of people with unrecognized anxiety disorders, namely those who have too little anxiety. Isaac Marks, the anxiety expert at the University of London, has coined the term "hypopho-bics" for such people. They don't complain and don't seek psychiatric treatment but instead end up in emergency rooms or fired from their jobs. As psychiatrists prescribe new antianxiety drugs with few side effects, we may create such conditions. For instance, one patient, shortly after starting on an antianxiety medication, impulsively told her husband that she wanted him to leave. He was very surprised but did. A week later she realized that she had three small children, a mortgage, no income, and no helpful relatives. A bit more anxiety would have inhibited such hasty action. Of course, no case is simple. This particular woman had long-standing marital dissatisfactions, and her emotional outburst might, in the long run, have left her better off. Her story illustrates one possible function of passions, as distinct from rational decisions. As suggested by Cornell economist Robert Frank, passions motivate actions that seem impulsive but may actually benefit the person in the long run.

NOVEL DANGERS

In the chapter on injuries, we described experiments that showed how monkeys' fear of snakes is "prepared." Most of our excessive fears are related to prepared fears of ancient dangers. Darkness, being away from home, and being the focus of a group's attention were once associated with dangers but now mainly cause unwanted fears. Agoraphobia, the fear of leaving home, develops in half of people who experience repeated panic attacks. Staying home seems senseless until you realize that most episodes of panic in the ancestral environment were probably caused by close encounters with predators or dangerous people. After a few such close calls, a wise person would try to stay home when possible, would venture out only with companions, and be ready to flee in panic at the least provocation: the exact symptoms of agoraphobia.

Do anxiety disorders, like many other diseases, result from novel stimuli not found in our ancestral environment? Not often. New dangers such as guns, drugs, radioactivity, and high-fat meals cause too little fear, not too much. In this sense we all have maladaptive hypophobias, but few of us seek psychiatric treatment to increase our fear. Some novel situations, especially flying and driving, do often cause phobias. In both cases, the fear has been prepared by eons of exposure to other dangers. Fear of flying has been prepared by the dangers associated with heights, dropping suddenly, loud noises, and being trapped in a small, enclosed place. The stimuli encountered in an automobile zooming along at sixty miles an hour are novel, but they too hark back to ancestral dangers associated with rapid movement, such as the rushing attack of a predator. Automobile accidents are so common and so dangerous that it is hard to say if fear of driving is beneficial or harmful.

The genetic contributions to anxiety disorders are substantial. Most people with panic disorders have a blood relative who has the same problem, and the search is on for the responsible genes. Will these genes turn out to result from mutant genes that have not been entirely selected out? Will they turn out to have other benefits? Or will we discover that genetic susceptibility to panic is simply one end of a normal distribution, like a tendency to develop a high fever with a cold or a tendency to vomit readily? When we find specific genes that predispose to panic and other anxiety disorders, we will still need to find out why those genes exist and persist.

SADNESS AND DEPRESSION

D epression sometimes seems like a modern plague. After motor vehicle accidents, suicide is the second leading cause of death of young adults in North America. Nearly 10 percent of young adults in the United States have experienced an episode of serious depression. Furthermore, the rates seem to have increased steadily in the past few decades, doubling every ten years in many industrial countries.

Depression may seem completely useless. Even apart from the risk of suicide, sitting all day morosely staring at the wall can't get you very far. A person with severe depression typically loses interest in everything—work, friends, food, even sex. It is as if the capacities for pleasure and initiative have been turned off. Some people cry spontaneously, but others are beyond tears. Some wake every morning at 4 A.M. and can't get back to sleep; others sleep for twelve or fourteen hours per day. Some have delusions that they are impoverished, stupid, ugly, or dying of cancer. Almost all have low self-esteem. It seems preposterous even to consider that there should be anything adaptive associated with such symptoms. And yet depression is so frequent, and so closely related to ordinary sadness, that we must begin by asking if depression arises from a basic abnormality or if it is a dysregulation of a normal capacity.

There are many reasons to think that the capacity for sadness is an adaptive trait. A universal capacity, it is reliably elicited by certain cues, notably those that indicate a loss. The characteristics of sadness are relatively consistent across diverse cultures. The hard part is figuring out how these characteristics can be useful. The utility of happiness is not difficult to understand. Happiness makes us outgoing and gives us initiative and perseverance. But sadness? Wouldn't we be better off without it? One test would be to find people who do not experience sadness and see if they experience any disadvantages. Or an investigator could use a drug that blocks normal sadness, a study that we fear may soon be conducted inadvertently on a massive scale as more and more people take the new psychoactive drugs. While we wait for such studies to be done, the characteristics of sadness and the situations that arouse it provide clues that may help us to discover its functions.

The losses that cause sadness are losses of reproductive resources. Whether of money, a mate, reputation, health, relatives, or friends, the loss is always of some resource that would have increased reproductive success through most of human evolution. How can a loss be an adaptive challenge, a situation that would benefit from a special state of preparation? A loss signals that you may have been doing something maladaptive. If sadness somehow changes our behavior so as to stop current losses or prevent future ones, this would be helpful indeed.

216

How can people behave differently after a loss in a way that increases fitness? First, you should stop what you are doing. Just as pain can make us let go of a hot potato, sadness motivates us to stop current activities that may be causing losses. Second, it would be wise to set aside the usual human tendency to optimism. Recent studies have found that most of us consistently overestimate our abilities and our effectiveness. This tendency to optimism helps us to succeed in social competition, where bluffing is routine, and also keeps us pursuing important strategies and relationships even at times when they are not paying off. After a loss, however, we must take off the rose-colored glasses in order to reassess our goals and strategies more objectively.

In addition to sudden losses, there are situations in which an essential resource is simply not available despite major expenditures and our best plans and efforts. Jobs end, friendships fade, marriages sour, and goals must be abandoned. At some point one must give up on a major life project in order to use the resources to start something else. Such giving up should not be done lightly. Quitting one's job shouldn't be done impulsively, because there are costs involved in retraining and starting at the bottom of another hierarchy. Likewise, it is foolish to casually give up any important relationship or life goal in which a major investment has already been made. So we don't usually make major life changes quickly. "Low mood" keeps us from jumping precipitously to escape temporary difficulties, but as difficulties continue and grow and our life's energies are progressively wasted, this emotion helps to disengage us from a hopeless enterprise so that we can consider alternatives. Therapists have long known that many depressions go away only after a person finally gives up some long-sought goal and turns his or her energies in another direction.

The capacity for high and low mood seems to be a mechanism for adjusting the allocation of resources as a function of the propitiousness of current opportunities. If there is little hope of payoff, it is best to sit tight rather than to waste energy. Real estate agents who enter the business during an economic downturn may be making a mistake. Students who are failing a course would sometimes do best to drop it and try another subject. Farmers who plant their fields during a drought may go broke. If, by contrast, we come upon a short-lived opportunity, then it may be best to make a major, intense effort, despite the possible risks, in order to have a chance at a big payoff. When a million dollars in cash fell out of the back of an armored car on the streets of Detroit, a few people who made an intense, brief effort profited nicely.

A better understanding of the functions of sadness will soon be essential. We are fast gaining the capacity to adjust mood as we choose. Each new generation of psychotropic drugs has increasing power and specificity with fewer side effects. Decades ago there was a hue and cry about "soma," the fictional drug that made people tolerate tedious lives in Aldous Huxley's *Brave New World*. Now that similar substances loom as a reality, strangely little is being said. Do people not realize how fast this train is moving? We certainly should try to relieve human suffering, but is it wise to eliminate normal low mood? Many people intuitively feel it is wrong to use drugs to change mood artificially, but they will have a hard time arguing against the use of nonaddicting drugs with few side effects. The only medical reason not to use such drugs is if they interfere with some useful capacity. Soon—very soon—people will be clamoring to know when sadness is useful and when it is not. An evolutionary approach provides a foundation for addressing these questions.

We are aware that this analysis is vastly oversimplified. People are not controlled by some internal calculator that crudely motivates them to maximize their reproductive success. Instead, people form deep, lifelong emotional attachments and experience loves and hates that shape their lives. They have religious beliefs that guide their behavior, and they have idiosyncratic goals and ambitions. They have networks of friends and relatives. Human reproductive resources are not like the squirrel's cache of nuts. They are, instead, constantly changing states of intricate social systems. All these complexities do not undercut our simple arguments; they just highlight the urgency of blazing the trail of functional understanding that the adaptationist program may provide for human emotions.

While some low mood is normal, some is clearly pathological. The causes of such pathology are complex. Genetic factors are important determinants of manic-depressive disorder, a condition in which mood swings wildly from the depths of depression to aggressive euphoria. Having one parent with manic-depressive disorder increases your risk of that disorder by a factor of 5, and having two parents increases it by a factor of 10 to a likelihood of nearly 30 percent. These genes are not rare—manic-depressive illness occurs in 1 out of 200 people. Our next, by now familiar, question is, Why are these genes maintained in the gene pool? The answer is equally familiar: They probably offer some advantage, either in certain circumstances or in combination with certain other genes. A study by Nancy Andreasen, professor of psychiatry at the University of Iowa, found that 80 percent of the faculty at the renowned Iowa Writer's Workshop had experienced some kind of mood disorder. Is creativity a benefit of the genes that cause depression? The disease wreaks havoc in some individual lives, but the genes that cause it seem nonetheless to offer a fitness advantage either to some people with the disorder or to other people in whom the gene does not cause the disorder but has other, beneficial effects.

John Hartung, an evolutionary researcher at the State University of New York, has suggested that depression is common in people whose abilities threaten their superiors. If a person with lower status demonstrates his or her full abilities, this is likely to bring attack from the more powerful superior. The best protection, Hartung suggests, is to conceal your abilities and to deceive yourself about them so as to more readily conceal your ambitions. This could well explain some otherwise mysterious cases of low self-esteem in successful people. Hartung's theory reminds us of the complexity of human emotions.

Another major effort to understand mood has come from a group of researchers who are pursuing British psychiatrist John Price's theory of the role of mood in human status hierarchies. They have argued that depression often results when a person is unable to win a hierarchy battle and yet refuses to yield to the more powerful person. They suggest that depression is an involuntary signal of submissiveness that decreases the likelihood of attacks by dominants. In case studies they describe how submitting voluntarily can end depression.

UCLA researchers Michael Raleigh and Michael McGuire have found a brain mechanism that connects mood and status. In studies of vervet monkeys, they found that the highest-ranking (alpha) male in each group had levels of a neurotransmitter (serotonin) that were twice as high as those of other males. When these "alpha" males lost their position, their serotonin levels immediately fell and they huddled and rocked and refused food, looking for all the world like depressed humans. These behaviors were prevented by the administration of antidepressants, such as Prozac, that raise serotonin levels. Even more astounding, if the researchers removed the alpha male from a group and gave antidepressants to some other randomly chosen male, that individual became the new alpha male in every instance. These studies suggest that the serotonin system may function, in part, to mediate status hierarchies and that some low mood may be a normal part of status competitions. If this is so, one cannot help but wonder what will happen in large corporations as more and more depressed employees start taking antidepressants.

Still another approach to understanding depression is based on the increase of the state that occurs when the amount of daylight decreases in the fall. The large number of people affected with this seasonal affective disorder (SAD) and its strong association with cold climates have suggested to many researchers that low mood may be a variant or remnant of a hibernation response in some remote ancestor. The preponderance of women with SAD has suggested that the response may somehow regulate reproduction.

Are there novel aspects of our modern environment that make depression and suicide more likely? While every age seems to have believed that people are not as happy as they were in earlier times, some recent evidence suggests that we may actually be in an epidemic of depression. A team of distinguished investigators looked at data from 39,000 people in nine different studies carried out in five diverse areas of the world and found that young people in each country are far more likely than their elders to have experienced an episode of major depression. Furthermore, the rates were higher in societies with higher degrees of economic development. Much remains to be done to confirm this finding, but it justifies an intense study of novel aspects of modern life that might contribute to dramatic increases in depression. We will mention only two: mass communications and the disintegration of communities.

Mass communications, especially television and movies, effectively make us all one competitive group even as they destroy our more intimate social networks. Competition is no longer within a group of fifty or a hundred relatives and close associates, but among five billion people. You may be the best tennis player at your club, but you are probably not the best in your city and are almost certainly not the best in your country or planet. People turn almost every activity into a competition, whether it be running, singing, fishing, sailing, seducing, painting, or even bird watching. In the ancestral environment you would have had a good chance at being best at something. Even if you were not the best, your group would likely value your skills. Now we all compete with those who are the best in all the world.

Watching these successful people on television arouses envy. Envy probably was useful to motivate our ancestors to strive for what others could obtain. Now few of us can achieve the goals envy sets for us, and none of us can achieve the fantasy lives we see on television. The beautiful, handsome, rich, kind, loving, brave, wise, creative, powerful, brilliant heroes we see on the screen are out of this world. Our own wives and husbands, fathers and mothers, sons and daughters can seem profoundly inadequate by comparison. So we are dissatisfied with them and even more dissatisfied with ourselves. Extensive studies by psychologist Douglas Kenrick have shown that after being exposed to photos or stories about desirable potential mates, people decrease their ratings of commitment to their current partners.

Our new technology also dissolves supportive social groups. For members of our socially oriented species, the worst punishment is solitary confinement, but many modern, anonymous groups are not much better. They often consist mostly of competitors with only an occasional comrade and no blood relatives. Extended families disintegrate as individuals scatter to pursue their economic goals. Even the nuclear family, that last remnant of social stability, seems doomed, with more than half of all marriages now ending in divorce and more and more children being born to single women.

We have a primal need for a secure place in a supportive group. Lacking family, we turn elsewhere to meet this need. More and more people have their social base in groups of friends, twelve-step programs such as Alcoholics Anonymous, support groups of all kinds, or psychotherapy. Many people turn to religion in part because of the group it provides. Some people advocate "family values" in hopes of preserving a threatened but cherished way of life. Most of us want most of all to be loved by someone who cares about us for ourselves, not for what we can do for them. For many, the search is bitter and fruitless.

Lack of Attachment

Pre-evolutionary theories, both psychoanalytic and behavioral, explained the bond between mother and child as the result of feeding and caretaking. Primatologist Harry Harlow began to challenge these theories with studies of monkeys at the University of Wisconsin in the early 1950s. Infant monkeys were separated from their mothers and provided with two surrogate mothers, one a wire form with a baby bottle full of milk, the other a soft cloth-covered form without a bottle. Although infants got milk from the wire mother, it was the cloth surrogate they clung to, screaming if it was removed. Harlow concluded that there must be a special mechanism that evolved to facilitate the bonding of mother and infant. Inspired by Rene Spitz's studies of the social inadequacy of children raised in orphanages, Harlow next raised monkey infants in isolation. Such monkeys never became normal. They could not get along with other monkeys, had great difficulty in mating, and neglected or attacked any babies they had.

John Bowlby, an English psychiatrist, attended seminars with biologist Julian Huxley in 1951 and was inspired to read the imprinting experiments done by Nobel Prize–winning ethologist Konrad Lorenz. During a very specific critical period early in life, baby goslings imprint on their mothers or any other appropriate-sized moving object they encounter. Konrad Lorenz's boots were sufficiently similar, and many photos show him being trailed by a line of goslings. Bowlby wondered if many of his patients' difficulties were sequelae of problems with early attachment. As he looked at their first relationships, he found problems everywhere. Some had mothers who had never wanted them, others had mothers who were too depressed to respond to smiles and coos. Many had heard their mothers threaten to kill themselves and had grown up under this specter. People's early difficulties matched the problems they experienced as adults. They could not trust people, they expected to be rejected, and they felt they had to please people or they would be abandoned. Bowlby perceptively recognized that some of the clinging and withdrawal behavior of neglected babies might be adaptive attempts to engage the mother. Instead of criticizing patients for being "dependent," he recognized that they were trying to protect themselves from a feared separation.

Psychologist Mary Ainsworth and her colleagues did the controlled studies that brought Bowlby's theories to mainstream psychology. She put young children into a room and observed their behavior when the mother left and later returned. On the basis of this "strange situation" test, she classified babies into those who were securely attached and those who were anxiously attached or who avoided their mothers on reunion. Which group the child fit into strongly predicted many other characteristics from group-play patterns to personality characteristics many years later. Much remains to be done to determine what the relationship is between attachment problems and adult psychopathology and how it relates to genetic factors. Psychiatrists should not forget that mothers provide not only early experiences for their children; they also provide genes. At present we have reason to believe that many problems adults have in getting along with other people may have their origins in problems with the first attachment.

CHILD ABUSE

C hild abuse seems to have become epidemic among us. How can this be? Why would we attack our own children, the vehicles of our reproductive success? Are some parents more likely to abuse than others? Canadian psychologists Martin Daly and Margo Wilson's evolutionary perspective led them to wonder if the presence or absence of a blood relationship between parents and children might predict the likelihood of child abuse. Because of the vagaries in the reporting of child abuse, they looked at an outcome that was easy to count and hard to hide—murders of children by their parents. The correlation was stronger than even they had dared to imagine. The risk of fatal child abuse for children living with one nongenetic parent is seventy times higher than it is for children living with both biological parents. This finding was not explainable by any tendency of families with stepparents to have more alcoholism, poverty, or mental illness. In several decades of research, no other risk factor has proved anywhere near as powerful in predicting child abuse. Many who have studied child abuse for decades never thought to look at the significance of kinship, but to evolutionists this was an obvious suspect.

Daly and Wilson were inspired, in part, by studies on infanticide in animals carried out by California anthropologist Sarah Hrdy and others. When Hrdy reported in 1977 that male languar monkeys routinely tried to kill the infants of females in a group they had just taken over from another male, no one wanted to believe her. She reported that the monkey mothers tried to protect their infants but often did not succeed. When they failed, nursing stopped, estrus came quickly, and the monkey mothers promptly mated with the males who had killed their infants. Hrdy noted that males who killed existing infants would increase their reproductive success because the cessation of nursing brought the females into estrus so they could become pregnant with the offspring of the new male sooner.

Subsequent field research has confirmed Hrdy's findings and extended them to many other species. Male lions kill existing cubs when they begin mating with new females. Among mice, the mere smell of a strange male often induces miscarriage—apparently an adaptation to prevent wasting investment on babies that are likely to be killed. Animals are inevitably designed to do whatever will increase the success of their genes, grotesque though the resulting behavior may seem.

The tendency for male animals to kill the offspring of other males in certain circumstances is an evolved adaptation. Is child abuse in humans in any way related? We had thought not, both because human males don't routinely take over a group of breeding females with young offspring and because many foster fathers are obviously capable of providing excellent care for children who are not their own. We had guessed that children are abused not because of an evolved adaptation but because a normal adaptation failed when one of the parents had too little early contact with the child to facilitate normal attachment. However, studies by anthropologist Mark Flinn in Trinidad have found that stepparents still treat their stepchildren more harshly than their natural children, regardless of the amount of early contact with the baby. More is involved in forming human attachments than merely spending time together. Much more research is needed to explore this murky intersection of biology and culture.

SCHIZOPHRENIA

The symptoms of schizophrenia, unlike those of anxiety and depression, are not a part of normal functioning. Hearing voices, thinking that others can read your mind, emotional numbness, bizarre beliefs, social withdrawal, and paranoia appear together as a syndrome not because they are parts of an evolved defense. It is more likely that one kind of brain damage can cause many malfunctions, just as heart damage can cause shortness of breath, chest pain, and swollen ankles. Schizophrenia disrupts the perceptual-cognitive-emotional-motivational system. This is another way of saying that we still don't know how to describe the higher levels of brain function.

Schizophrenia affects about 1 percent of the population in diverse societies worldwide. The notion that it is a disease of civilization seems to be incorrect, although there have recently been suggestions that the course of the disease is worse in modern societies. Compelling evidence suggests that susceptibility to schizophrenia depends on certain genes. Relatives of schizophrenics are several times more likely than other people to get the disease, even if they were raised by nonschizophrenic adoptive parents. If one identical twin has schizophrenia, the chance of the other getting it is about 50 percent, while the risk for a nonidentical twin is about 25 percent. There is also evidence that schizophrenia decreases reproductive success, especially in men.

These observations call up our standard question: What can account for the high incidence of genes that can decrease fitness? Selection against the genes that cause schizophrenia is strong enough that they should be far less common if their presence were due simply to mutation balanced by selection. Furthermore, the relatively uniform rates of schizophrenia suggest that the responsible genes did not arise recently but have been maintained for millennia. It appears that the genes that cause schizophrenia must somehow confer an advantage that balances the severe costs.

The most likely possibility is that these genes are advantageous in combination with certain other genes, or in certain environments, much in the way a single sickle-cell gene is advantageous even though having two such genes causes sickle-cell anemia. Or it might be that the genes that predispose to schizophrenia have other effects that offer a slight advantage in most people who have them, even though a small proportion develop the disease. A number of authors have speculated on the kinds of advantages that might accrue to people who have genes that predispose to schizophrenia: perhaps they increase creativity or sharpen a person's intuitions about what others are thinking. Perhaps they protect against some disease. Some have suggested that the tendency to suspiciousness itself may compensate somewhat for the disadvantages of schizophrenia. Evidence for these ideas remains scattered, but they are worth pursuing. Support is provided by evidence of high levels of accomplishment in relatives of schizophrenics who are not affected by the disease. This whole area is just beginning to be explored.

SLEEP DISORDERS

S leep, like so many other bodily capacities, commands our attention only when it goes awry, which it does for many people in many ways. For sleep, as for so many things, timing is often the crucial factor. Most sleep problems involve an inability to sleep at the proper time or a tendency to sleep at the wrong time. Insomnia affects more than 30 percent of the population and is the spur to a huge industry, from over-the-counter sleeping pills to specialized medical clinics. The people who suffer from daytime sleepiness are often the same ones who don't sleep well at night. Sleepiness is a bother when you are trying to read in the evening, a handicap after the alarm rings in the morning, and a positive danger if it happens while you are driving.

Then there are dreams and their disorders, nightmares and night terrors. Some people experience a kind of lack of coordination of the aspects of sleep and become conscious while still dreaming and unable to move, a frightening state indeed. People with narcolepsy slip suddenly into dreaming sleep in the midst of everyday activities, sometimes so swiftly that they fall and injure themselves. And then there are the people with sleep apnea, who intermittently stop breathing during sleep with resulting nighttime restlessness, daytime tiredness, and even brain damage. In order to understand these problems, we need to know more about the origins and functions of normal sleep.

Is sleep a trait that has been shaped by natural selection? There are several reasons to think so. First, the trait is widespread among animals and perhaps universal among vertebrates. In some animals that seem not to sleep, such as dolphins, one half of the brain in fact sleeps while the other stays awake, possibly because they must repeatedly swim to the surface to breathe. Second, all vertebrates seem to share the same sleep regulation mechanisms, with the center that controls dreaming sleep consistently located in the ancient parts of the brain. Third, the patterns of mammalian sleep, with its periods of rapid eye movement and rapid brain waves, are also shared with birds, whose evolution diverged from ours before the time of the dinosaurs. Fourth, the wide variation in the actual patterns of sleep, even in closely related mammals, suggests that whatever kind of sleeping was done by our most recent common ancestor could evolve rapidly to match the species' particular ecological niche. Finally, if deprived of sleep, all animals function poorly.

In order to better understand sleep difficulties, we would like to understand how the capacity and necessity for sleep increase fitness. One major contribution to the problem came in 1975 from British biologist Ray Meddis, who proposed that the amount and timing of our sleep are set by our potential for productive activity in different phases of the day-night cycle. As one reviewer of Meddis's book put it, our motivation to sleep at night arises from the desirability of staying off the streets. If there are special dangers in being abroad in the dark and little likelihood of positive accomplishment then, we are better off resting. This explains why humans and other animals benefit from a daily cycle of activity, but it does not explain why we sleep instead of just spending the night quietly awake, ready for any opportunities or dangers that may arise. It also does not explain why we have become so dependent on sleep that its lack makes us barely able to function.

Here is one possible perspective on the evolutionary origins of sleep. Imagine that some distant ancestor needed no sleep. If one line of its descendents had experienced greater dangers at one part of the day-night cycle (let's assume for simplicity that it was night) and greater opportunities during the day, then individuals who were inactive at night would have had a fitness advantage. As the species gradually came to confine its activity to the daylight hours, its nocturnal quiescence grew ever more prolonged and profound until it reliably spent many hours of every night inactive.

Given such a reliable daily period of inactivity, other evolutionary factors would be expected to act. It is unlikely that all needed cellular maintenance activities would proceed equally well whether an animal were awake or asleep. If some needed processes worked more efficiently when the brain was disengaged from its usual tasks, selection would act to delay them during the wakeful day and catch up during the night, thus favoring development of the state we recognize as sleep. In this way, as suggested in 1969 by Ian Oswald of Edinburgh University, some brain maintenance processes would be confined more and more to sleep and we would become more and more dependent on sleep. During this period, of course, it would be necessary for sleeping individuals to be quite safe, otherwise sleep would quickly have been selected against. Just as we became dependent on getting vitamin C from foods only because we could reliably get plenty of it, the steady availability of a period of safe rest was necessary before certain bodily maintenance mechanisms could be carried out only during sleep. One implication is that a search for metabolic processes confined to sleep, or taking place at a much greater rate during sleep, will provide insights on why we need to sleep. Indeed, brain scans have shown that protein synthesis is greatest during dreamless sleep and that mechanisms for synthesizing certain neurotransmitters can't keep up with daytime utilization and therefore must catch up at night. Furthermore, cell division is fastest in all tissues during sleep.

Once sleep was established for physiological repair, natural selection might well have relegated other functions to this period. Those most often suggested have been the memory-regulation functions. Researchers Allan Hobson and Robert McCarley have argued that dreaming sleep supports the physiology that consolidates learning. Francis Crick and Graeme Mitchison have evidence that dreaming sleep functions to purge unnecessary memories, much as we periodically discard unnecessary files from our computers. We won't consider these suggestions in detail but will only point out only that these are not necessarily mutually exclusive alternatives, nor are they at odds with Oswald's idea that sleep evolved as a period of tissue repair. None of this contradicts Meddis's observation that sleep regulates activity periods depending on the animal's ecology. Like other traits, sleep undoubtedly has many important functions. While each hypothesized function needs to be tested, support for one alternative provides evidence against another only if the functions are incompatible. Studies of sleep patterns in many different animal groups in relation to their ways of life and evolutionary relationship to one another could provide helpful evidence.

Now that we are seldom threatened by nocturnal predators such as tigers, and now that artificial light makes productive activity possible throughout the night, the need for regular sleep has become a great bother, especially when we fly across the world and our bodies insist on living according to our original time zone. Looking for the functions of sleep may well provide the knowledge we need to adapt it better to our present needs—or, at the very least, to make it possible to read in the evening without falling asleep and then to sleep soundly through the night despite our worries about the crises tomorrow might bring.

DREAMING

Dreaming has interested people since the dawn of history and no doubt through much of prehistory. In recent years, many theories have been proposed about the functions of dreams, from Freud's theory of dreams fulfilling forbidden wishes to Francis Crick's theory that dreams erase and reorganize memories. But the debate has been so inconclusive that some current major authorities, like Harvard's Allan Hobson, can still argue that dreams may have no specific function but are mainly epiphenomena of brain activities. This seems unlikely to us, given the simple observation that deprivation of dreaming sleep causes severe psychopathology. For instance, cats kept on tiny islands in a pool were able to sleep, but the loss of muscle tone that accompanies dreaming sleep slipped them into the water and woke them. Such deprivation of dreaming sleep made these unfortunate cats wild and hypersexual and shortened their lives.

Even without delineating the function of dreams, an evolutionary approach can contribute to their understanding. Donald Symons, an evolutionary anthropologist at the University of California (Santa Barbara), recently proposed that there are, for evolutionary reasons, serious constraints on the stimuli we experience in dreams. While individual sleep behavior varies enormously, we tend, in dreams, to experience a wealth of our own actions and of sights but very little sound, smell, or mechanical stimulation. We can dream about doing things without actually moving because our motor nerves are paralyzed when we are in the kind of sleep that permits dreaming. We remember what people in dreams look like and what they tell us, but we do not remember as easily what their voices sounded like. We may remember enjoying a dreamworld glass of wine, but we often cannot recall its bouquet. We can dream that someone strikes us but may not remember what it felt like.

The reason for these constraints, Symons suggests, is that they were required by Stone Age realities. We could afford visual hallucinations, because closed eyes made sight useless; it was too dark for effective vision anyhow. By contrast, a cry of alarm, the smell of a tiger, or the panicky grasp of a child were important cues that required unimpaired vigilance of our senses of hearing, smell, and touch. Some species sleep with their eyes open, but we sleep with our ears open: we cannot let our dreams distract us from important sounds. Symon's theory explains some of the peculiarities of dreaming (and predicts some not yet noticed), and it will stand or fall according to how well its expectations conform to actual findings on the sensory composition of dreams. So far it seems to account for most of the available evidence.

THE FUTURE OF PSYCHIATRY

Psychiatry has recently emulated the rest of medicine by devising clear (if somewhat arbitrary) diagnostic categories, reliable methods of measuring symptoms, and standard requirements for experimental design and data analysis. Psychiatric research is now just as quantitative as that in the rest of medicine. Has all this apparent rigor brought psychiatry acceptance as just another medical specialty like neurology, cardiology, or endocrinology? Hardly. The research findings are solid, but they are not connected in any coherent theory. In its attempt to emulate other medical research by searching for the molecular mechanisms of disease, psychiatry has ironically deprived itself of precisely the concepts that provide the tacit foundation for the rest of medical research. By trying to find the flaws that cause disease without understanding normal functions of the mechanisms, psychiatry puts the cart before the horse.

Research on the anxiety disorders exemplifies the problem. Psychiatrists now divide anxiety disorders into nine subtypes, and many researchers treat each as a separate disease, investigating its epidemiology, genetics, brain chemistry, and response to treatments. The difficulty is, of course, that anxiety is not itself a disease but a defense. To appreciate the problems this creates, imagine what would happen if doctors of internal medicine studied cough the way modern psychiatrists study anxiety. First, internists would define "cough disorder" and create objective criteria for diagnosis. Perhaps the criteria would say you have cough disorder if you cough more than twice per hour over a two-day period or have a coughing bout that lasts more than two minutes. Then researchers would look for subtypes of cough disorder based on factor-analytic studies of clinical characteristics, genetics, epidemiology, and response to treatment. They might discover specific subtypes of cough disorders such as mild cough associated with runny nose and fever, cough associated with allergies and pollen exposure, cough associated with smoking, and cough that usually leads to death. Next, they would investigate the causes of these subtypes of cough disorder by studying abnormalities of neural mechanisms in people with cough disorders. The discovery that cough is associated with increased activity in the nerves that cause the chest muscles to contract would stimulate much speculation about what neurophysiological mechanisms could make these nerves overly active. The discovery of a cough-control center in the brain would give rise to another set of ideas as to how abnormalities in this center might cause cough. The knowledge that codeine stops cough would lead other scientists to investigate the possibility that cough results from deficiencies in the body's codeinelike substances.

Such a plan of research is obviously ludicrous, but we recognize its folly only because we know that cough is useful. Because we know that cough is a defense, we look for the causes of cough not in the nerves and muscles that generate a cough, or even in the brain mechanisms that regulate cough, but instead in the situations and stimuli that normally arouse the protective cough reflex. While some rare cases of cough may be caused by abnormalities of the cough-regulation mechanisms, the vast majority are adaptive responses that expel foreign matter from the respiratory tract. Only after searching for such a natural stimulus does a physician consider the possibility that the cough-regulation mechanism itself might be awry.

Many psychiatrists have studied individual differences in susceptibility to anxiety with the worthy goal of helping the many people who experience panic, tension, fear, and sleeplessness throughout their lives. Nonetheless, this approach fosters much confusion. What if research on cough were to focus on those individuals who have a lifelong tendency to cough in response to the least stimulus? Such people would be told they have a cough disorder. Soon there would be campaigns to identify people predisposed to cough disorder in order to find the genes that cause this abnormality in the cough-regulation mechanism. There undoubtedly are people with a genetic susceptibility to ready coughing, but studying them would tell us little about the cause of most coughs.

There are limits to this analogy. Anxiety is much more complicated than cough, its functions are less obvious, and it varies much more from individual to individual. More important, the cues that arouse anxiety are far less tangible than those that arouse cough. Cough is caused by foreign material in the respiratory tract, while anxiety is aroused by diverse cues processed by the mind in mysterious ways. The most obvious anxiety cues are images of dangerous objects or stimuli that have been paired with pain or some other noxious stimulus. Most clinical anxiety is aroused, however, by complex cues that require subtle interpretation. If, for example, the boss doesn't greet you, you are not invited to a meeting, and a friend avoids you on a day when layoff notices are to be distributed, you may feel serious apprehension. If it is your birthday, however, and you suspect a surprise party may be in the works, the same stimuli will arouse a very different reaction. This example only begins to tap the complexity of the mental systems that regulate anxiety. Many wishes and feelings never make it to consciousness but nonetheless cause anxiety. The woman whose panic attacks started when she began an affair insisted that the two were unrelated. Just because many of the cues that cause anxiety are hard to identify does not mean that they are not there, and it certainly does not mean that the anxiety they cause is useless or a product of abnormal brain mechanisms.

Conversely, just because much anxiety is normal, that does not mean it is all useful. Furthermore, many anxiety disorders *are* caused by genetic predispositions. We don't yet know whether these are best understood as genetic defects or normal variations. Certainly, the kinds and dangerousness of various threats vary considerably from one generation to the next, and this should maintain considerable genetic variation in the anxiety-regulation mechanisms.

If psychiatry stays on its current course, it will be left treating only those disorders caused by demonstrable brain defects, while the pains and suffering of everyday life will be left to other clinicians. This would be unfortunate for patients as well as psychiatrists. The rest of medicine treats normal defensive reactions; why shouldn't psychiatry do the same? In this as well as other ways, an evolutionary view is psychiatry's route to genuine integration with the rest of medicine. An intensive effort to understand the functions of the emotions and how they are normally regulated would provide, for psychiatry, something comparable to what physiology provides for the rest of medicine. It would provide a framework in which pathopsychology could be studied like pathophysiology, so that we can understand what has gone wrong with the normal functioning of bodily systems. There is every expectation that an evolutionary approach will bring the study of mental disorders back to the fold of medicine, relying not on a crude "medical model" of emotional problems but on the same Darwinian approach that is so useful in the rest of medicine.

15

THE EVOLUTION OF MEDICINE

Nothing in biology makes sense except in the light
of evolution.
———Theodosius Dobzhansky, 1973

You are crossing a heath on a well-worn path when a flash
of early sunlight reflects from something lying over by an
older trail. You follow the gleam to its source, brush away
some dirt, and discover an old-fashioned gold pocket
watch. Perhaps it is the same old watch that people have been finding
for two centuries, but some details have been overlooked.

Its perfection still elicits wonder. The seam around the case is all
but invisible; the crystal is symmetrical and gleaming; the chain is
made of exquisitely miniature gold links. The face has numerals
sharply etched around the logo of the Lifetime Watch Company. But
even as you admire the watchmaker's skill, the light reveals some surprising imperfections. The crystal is laced with slight distortions.
And the chain, though beautiful and flexible, is thin and broken, thus
explaining why the watch is here and not in a pocket. A notch in the
seam is perfectly shaped for a thumbnail but large enough for dirt
and water to enter easily. Odd, these flaws. You open the back, and
the exquisite mechanism again inspires awe. How could anyone have
designed, much less constructed, so many perfectly cut gears of rust-
proof brass, the hairlike spring of steel, the balance wheel suspended

by minuscule jewels? But when you try to set the watch, the knob is so tiny you can barely grasp it and a dozen twirls advance the hands only a single hour. You shake the watch. It ticks for five seconds, then is stopped by flakes of rust from that steel spring. What an odd device this is! So perfect in many respects, in others makeshift at best. How could the creator of such a masterpiece have allowed so many obvious flaws? Inside the case is an inscription in tiny letters. You take out your magnifying glass and read:

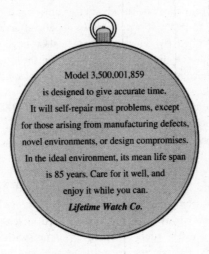

Model 3,500,001,859
is designed to give accurate time.
It will self-repair most problems, except
for those arising from manufacturing defects,
novel environments, or design compromises.
In the ideal environment, its mean life span
is 85 years. Care for it well, and
enjoy it while you can.
Lifetime Watch Co.

OVERVIEW OF CAUSES OF DISEASE

We now return to where we began, to a seeming incongruity at the core of medicine. Despite their exquisite design, our bodies have crude flaws. Despite our multiple defenses, we have a thousand vulnerabilities. Despite their capabilities for rapid and precise repairs, our bodies inevitably deteriorate and eventually fail. Before Darwin, physicians could only wonder at the incongruity of it all, perhaps with the hope that our bodies are part of an unfathomable divine plan, or with the suspicion that they are some cosmic prank. Ever since Darwin, the incongruity has often mistakenly been attributed to the supposed weakness or capriciousness of natural selection. In the light of modern Darwinism, however, the incongruity unfolds into a sharply blocked tapestry with a place for each of several distinct causes of disease.

Why isn't the body more reliable? Why is there disease at all? As we have seen, the reasons are remarkably few. First, there are genes that make us vulnerable to disease. Some—though fewer than has been thought—are defectives continually arising from new mutations but kept scarce by natural selection. Other genes cannot be eliminated because they cause no disadvantages until it is too late in life for them to affect fitness. Most deleterious genetic effects, however, are actively maintained by selection because they have unappreciated benefits that outweigh their costs. Some of these are maintained because of heterozygote advantage; some are selected because they increase their own frequency, despite creating a disadvantage for the individual who bears them; some are genetic quirks that have adverse effects only when they interact with a novel environmental factor.

Second, disease results from exposure to novel factors that were not present in the environment in which we evolved. Given enough time, the body can adapt to almost anything, but the ten thousand years since the beginnings of civilization are not nearly enough time, and we suffer accordingly. Infectious agents evolve so fast that our defenses are always a step behind. Third, disease results from design compromises, such as upright posture with its associated back problems. Fourth, we are not the only species with adaptations produced and maintained by natural selection, which works just as hard for pathogens trying to eat us and the organisms we want to eat. In conflicts with these organisms, as in baseball, you can't win 'em all. Finally, disease results from unfortunate historical legacies. If the organism had been designed with the possibility of fresh starts and major changes, there would be better ways of preventing many diseases. Alas, every successive generation of the human body must function well, with no chance to go back and start afresh.

The human body turns out to be both fragile and robust. Like all products of organic evolution, it is a bundle of compromises, each of which offers an advantage, but often at the price of susceptibility to disease. These susceptibilities cannot be eliminated by any duration of natural selection, for it is the very power of natural selection that created them.

RESEARCH

Many questions confront the infant enterprise of Darwinian medicine. What is its long-range goal? How should we go about analyzing a disease from an evolutionary viewpoint? How should hypotheses be formulated and tested? Who will pay for this research? Who will do the research and in what academic departments or other agencies? Why has it taken so long to get this enterprise started?

We begin with the long-range goal. What will medical textbooks look like when evolutionary studies of disease are well established? Current textbooks summarize what is known about a disorder under traditional headings: signs and symptoms of the disease, laboratory findings, differential diagnosis, course, complications, epidemiology, etiology, pathophysiology, treatment, and outcome. Such descriptions fall one category short. A comprehensive discussion of a disease must also provide an evolutionary explanation. While some current textbooks have a sentence or two about the advantages of the sickle-cell gene or the benefits of cough or fever, none of them systematically addresses the evolutionary forces acting on genes that cause disease, the novel aspects of environment that cause disease, or the details of the host-parasite arms race. Every textbook description of a disease should have, in our opinion, a section devoted to its evolutionary aspects. This section should address the following questions:

1. Which aspects of the syndrome are direct manifestations of the disease, and which are actually defenses?

2. If the disease has a genetic component, why do the responsible genes persist?

3. Do novel environmental factors contribute to the disease?

4. If the disease is related to infection, which aspects of the disease benefit the host, which benefit the pathogen, and which benefit neither? What strategies does the pathogen use to outflank our defenses, and what special defenses do we have against these strategies?

5. What design compromises or historical legacies account for our susceptibility to this disease?

Such questions immediately suggest important but neglected research on many diseases. Even the common cold offers many opportunities. What are the effects of taking or not taking aspirin? What are the effects of using nasal inhalers or decongestant medication? To use the categories of Chapter 3, is rhinorrhea (runny nose) a defense, a means the virus uses to spread itself, or both? For the most part, these projects have yet to be undertaken despite their conceptual simplicity and their obvious practical implications for us all.

Take something far more chronic and complicated, plantar fasciitis. More often known as heel spurs, this common disorder causes intense pain on the inside edge of the heel, especially first thing in the morning. The proximate cause is inflammation at the point where the heel attaches to the plantar fascia, a band of tough tissue that connects the front and rear of the foot like the string on a bow, supporting the arch of the foot. With every footstep it stretches, bearing the weight of the body thousands of times every day. Why does this fascia fail so often? The easy answer is that natural selection cannot shape a tissue strong enough to do the job—but by now this explanation should be suspect. Somewhat more plausible is the possibility that we began walking on two feet so recently that there has not been enough time for natural selection to strengthen the fascia sufficiently. The problem with this explanation is that plantar fasciitis is common and crippling. Like nearsightedness, it would, in the natural environment, so drastically decrease fitness that it would be strongly selected against. Some experts say plantar fasciitis arises in people who walk with their toes pointed out, a conformation that puts increased stress on the tissue. But then why do we walk that way? Is it the modern habit of wearing shoes? But many people who have never worn shoes also walk with their toes pointed outward.

Two clues suggest that plantar fasciitis may result from environmental novelty. First, exercises that stretch the plantar fascia to make it longer and more resilient are effective in relieving the problem. Second, many of us do something hunter-gatherers don't: we sit in chairs all day. Most hunter-gatherers walk for hours each day, instead of compressing their exercise into an efficient aerobic workout. When they aren't walking, they don't use chairs, they squat, a position that steadily stretches the plantar fascia. No plantar fasciitis and physical therapy for them, just squatting and walking for hours each day. This hypothesis, that plantar fasciitis results from prolonged sitting that allows the fascia to contract and that the disorder can be prevented and relieved by squatting and other stretching of the fascia, can readily be tested with epidemiological data and straightforward treatment studies.

Another good challenge for Darwinian medicine is the current controversy about whether it is wise to take antioxidants such as vitamin C, vitamin E, and beta-carotene. Folklore has long credited these agents with reducing heart disease, cancer, and even the effects of aging. Controlled studies are increasingly supporting these claims, especially for the prevention of atherosclerosis, although a major study in 1994 reported that beta-carotene appeared to *increase* the risk of cancer in some people. The agents are still deemed controversial, and many physicians studying them recommend caution until larger studies can assess their risks as well as benefits. We agree with this general conservatism but hope that an evolutionary view can speed the process. Earlier in this book we noted that natural selection seems to have resulted in high levels of several of the body's own antioxidants even though they cause disease. Uric acid levels are higher in species that live longer and are so high in humans that we are susceptible to gout. It appears that natural selection has acted to increase the human levels of uric acid, superoxide dismutase, and perhaps bilirubin and other substances as well, because they are antioxidants that slow some effects of aging in a species that has greatly increased its life span in just the past few hundred thousand years.

Why doesn't the body have antioxidant levels that are already optimal? It is possible that our antiaging mechanisms are still catching up with the recent increase in our life span. It is also possible that the costs of high levels of antioxidants (perhaps decreases in our resistance to infection or toxins?) have restricted them to levels that were optimal for a normal Stone Age lifetime of thirty or forty years. These possibilities suggest that adding extra antioxidants to the diet may have benefits that exceed the costs. In contrast to the many cases in which an evolutionary view argues against excessive intervention, here it supports the active pursuit of strategies that may prevent some effects of aging. A major part of such studies should be a search for other antioxidants in the body and an assessment of their costs and benefits. It would be interesting to see if people with high uric acid levels have costs other than gout and whether they show fewer signs of aging than other people. It will also be important to look for similar costs and benefits in our primate relatives. With this knowledge we will be in a better position to decide who will benefit from taking antioxidants and what the side effects might be.

This book contains suggestions for dozens of studies, many of which seem to us to be fine topics for Ph.D. theses and some of which offer challenges enough for a whole career. Pursuing them will be difficult, however, because no government agency presently supports such projects. Existing funding committees are reluctant to provide support because their mandate is to provide funds to study the proximate mechanisms and treatment of particular diseases. Furthermore, few members of such committees know anything about the formulation or testing of evolutionary ideas, and some are likely to have misgivings based on fundamental misconceptions about the scientific status of evolutionary hypotheses. The system used to assign funding priorities ensures that even a few people with such misgivings can eliminate the chances of funding.

Asking biochemists or epidemiologists to judge proposals to test evolutionary hypotheses is like asking mineral chemists to judge proposals on continental drift. Darwinian medicine needs its own funding panels staffed by reviewers who know the concepts and methods of evolutionary biology. Realistically, the prospects are poor for major government funding soon. The best hope for rapid growth of the field lies in the vision of private donors or foundations that could create institutes to support the development of Darwinian medicine. Even moderate support of this sort could quickly change the course of medicine, just as prior investments in biochemical and genetic research are now transforming our lives. As René Dubos noted in 1965:

> In many ways, the present situation of organismic biology and especially of environmental medicine is very similar to that of the physicochemical sciences related to medicine around 1900. At that time there was no place in the United States dedicated to the pursuit of physicochemical biology, and the scholars who were interested in this field were treated as second-class citizens in the medical community. Fortunately, a few philanthropists were made aware of this situation, and they endowed new kinds of research facilities to change the trend. The Rockefeller Institute is probably the most typical example of a conscious and successful attempt to provide a basis of physicochemical knowledge for the art of medicine. . . . Organismic and especially environ-

mental medicine constitute today virgin territories even less developed than was physicochemical biology 50 years ago. They will remain undeveloped unless a systematic effort is made to give them academic recognition and to provide adequate facilities for their exploration.

WHY DID IT TAKE SO LONG?

W hy has it taken more than a hundred years to apply Darwin's theory systematically to disease? Historians of science will eventually address this question, but from this close perspective several explanations seem likely: the supposed difficulty in formulating and testing evolutionary hypotheses about disease, the recency of some advances in evolutionary biology, and some peculiarities of the field of medicine.

Biologists have long tried to figure out the evolutionary origins and functions for organismic characteristics, but it has taken a surprisingly long time to realize that this enterprise is fundamentally different from trying to figure out the structure of organisms and how they work. Harvard biologist Ernst Mayr, in *The Growth of Biological Thought*, traces the parallel development of the two biologies. Medicine, while at the forefront of proximate biology, has been curiously late in addressing evolutionary questions. This is, no doubt, in part simply because the questions and goals are so different. It takes a wrenching shift to stop asking why an individual has a particular disease and to ask instead what characteristics of a species make all of its members susceptible to that disease. It has seemed a bit odd until now even to ask how something maladaptive like disease might have been shaped by natural selection. Furthermore, medicine is a practical enterprise, and it hasn't been immediately obvious how evolutionary explanations might help us prevent or treat disease. We hope this book convinces many people that seeking evolutionary explanations for disease is both possible and of substantial practical value.

If we are to assign blame for the tardiness of medicine in making use of relevant ideas in evolutionary biology, it rests as much with evolutionary biologists as with the medical profession. It took evolutionists an inexcusably long time to formulate the relevant ideas.

Given the powerful insights of Darwin, Wallace, and a few others in the middle of the nineteenth century, and the Mendelian revolution in genetics in the early years of the twentieth, why was it not until Fisher's book of 1930 that we had the first fruitful idea about why the number of boys and girls born is nearly equal? Why was it not until Medawar's midcentury work that we had any idea why there is such a thing as senescence? Why was it not until Hamilton's publications in 1964 that there was any realization that kinship would have some relevance to evolution? Why was it not until the 1970s and 1980s that we had useful ideas on how parasites and hosts, or plants and herbivores, influence each other's evolution? We believe that the answers to these and related questions will be found in a persistent antipathy to evolutionary ideas in general and to adaptation and natural selection in particular (even among some biologists). Meanwhile, we will simply note that medical researchers can hardly be blamed for failing to use the ideas of other sorts of scientists before those scientists developed them.

Medical scientists may also hesitate to consider functional hypotheses because of their indoctrination in the experimental method. Most of them were taught early, firmly—and wrongly—that science progresses only by means of experiment. But many scientific advances begin with a theory, and much testing of hypotheses does not rely on the experimental method. Geology, for instance, cannot replay the history of the earth, but it nonetheless can reach firm conclusions about how basins and ranges got that way. Like evolutionary hypotheses, geological hypotheses are tested by explaining available evidence and by predicting new findings that have not yet been sought in the existing record.

Finally, medicine, like other branches of science, is especially wary of ideas that in any way resemble recently overcome mistakes. Medicine fought for years to exclude vitalism, the idea that organisms were imbued with a mysterious "life force," so it continues to attack anything that is even vaguely similar. Likewise, *teleology* of a naive and erroneous sort keeps reappearing and must be expelled. Many people recollect from freshman philosophy class that teleology is the mistake of trying to explain something on the basis of its purpose or goal. This admonition is wise if it establishes an awareness that future conditions cannot influence the present. It is unwise if it also implies that present plans for the future cannot affect present processes and, through them, future conditions. Present plans may include printed recipes for baking cakes or the information in the DNA of bird's eggs. Functional explanations in biology imply not future influences on the present but a prolonged cycling of reproduction and selection. A bird embryo develops wing rudiments in the egg because earlier individuals that failed to do so left no descendants. Adult birds lay eggs in which embryos develop wing rudiments for the same reason. In this sense, a bird's wing rudiments are preparation for its future but are caused by its past history. Evolutionary explanations based on a trait's function do not imply that evolution involves any consciousness, active planning, or goal-directedness. While medicine is wise to be on guard against sliding back into discredited teleological reasoning, this wariness has prevented it from taking full advantage of the solid advances in mainstream evolutionary science. Through its efforts to keep from being dragged back, medicine has, paradoxically, been left behind.

MEDICAL EDUCATION

Medical education is similarly in trouble because of trying to guard against the old mistakes. The origins of its current quandary lie in the solution to a previous one. Early in this century, the Carnegie Foundation sponsored an extensive investigation of medical education by Abraham Flexner. In his cross-country travels, he reported a haphazard system of medical apprenticeship in which physicians, good and bad, took on assistants who, one way or another, learned something about medicine. Doctors' formal study of basic science was sporadic, and even their knowledge of basic anatomy and physiology was inconsistent. The Flexner report, published in 1910, formed the basis of new accreditation standards that required medical schools to provide future physicians with a foundation in basic science.

On this count, medical schools have far exceeded Flexner's hopes. In fact, one wonders what Flexner would say if he could see today's medical curricula. Now medical students are not only exposed to basic sciences, they are inundated with the latest advances by teachers who are subspecialist basic science researchers. At curriculum meetings in every medical school there are battles for students' time and minds. The microbiologists want more lab time, the anatomists want more too. The pathologists feel they cannot possibly fit their material into a mere forty hours of lecture. The pharmacologists say they will continue flunking 30 percent of the class until they get enough time to cover all the new drugs. The epidemiologists and biochemists and physiologists and psychiatrists and neuroscience experts all want more time, and certainly the students must keep up with the latest advances in genetics. Then they need to learn enough statistics and scientific methodology to be able to read the research literature. And they must somehow learn, before they start their work on the wards, how to talk with patients, how to do a physical exam, how to write up a patient report, how to draw blood, do a culture, a spinal tap, a Pap smear, measure eyeball pressure, examine urine and blood, and, and . . . The amounts of knowledge and the lists of tasks are overwhelming, but all must be completed in the first two years of medical school.

How is all this possible? It isn't. Why set impossible expectations? In part because we naively want our physicians to know everything. Another reason, however, is that no one person is in charge. When a committee decides on the class schedule and every basic science wants more time, the solution is to go on increasing the total amount of class time. Thirty or more hours each week in class is not unusual. After that, the students go home to study their textbooks and notes.

One might think that students' complaints would lead to reform, but decades of polite complaints changed little. It was technology that finally precipitated some change, technology in the form of the photocopy machine. Instead of going to class, students hire one person to take notes for each lecture, then all of them receive copies. It turns out to be a better survival strategy to stay home and study the notes than to go to class. When only twenty students attend a class for two hundred, professors hit the roof and curriculum reform is born. New attempts are being made, under the strong leadership of some deans, to cut back on the hours, reduce the amount of material, find new ways to transmit it. If these efforts succeed, it will be wonderful indeed.

Such efforts might even make room for Darwinian medicine, except that there are no Departments of Evolutionary Medicine to advocate inclusion of this material and few medical faculty members who know the material and want to teach it. It will take time and further leadership from medical school deans to make room in the medical curriculum for an introduction to the basic science of evolution and its applications in medicine. When evolution is included, it will give students not only a new perspective on disease but also an integrating framework on which to hang a million otherwise arbitrary facts. Darwinian medicine could bring intellectual coherence to the chaotic enterprise of medical education.

CLINICAL IMPLICATIONS

While many clinical implications of an evolutionary view await future research, others can immediately transform the way patients and doctors see disease. Let us listen in as first a pre-Darwinian and then a post-Darwinian physician talk to a patient about gout.

"So, Doctor, it is gout that has my big toe flaming, is it? What causes gout?"

"Gout is caused by crystals of uric acid in the joint fluid. I expect you can imagine only too well how some gritty crystals could make a joint painful."

"So why do I have it and you don't?"

"Some people have high levels of uric acid in their systems, probably because of some combination of genes and diet."

"So why isn't the body designed better? You would think there would be some system to keep uric acid levels lower."

"Well, we can't expect the body to be perfect, now, can we?"

At this point our pre-Darwinian physician gives up on science and dodges the question, implying that such "why" questions need not be taken seriously. Most likely, he or she doesn't recognize the distinction between proximate and evolutionary explanations, to say nothing of the importance and legitimacy of evolutionary explanations for disease.

The Darwinian physician gives a different answer, one closer to what the patient wanted and was entitled to.

"That's a good question. It turns out that human uric acid levels are much higher than those of other primates and that uric acid levels in a species are correlated with its life span. The longer-lived the species, the higher the uric acid levels. It seems that uric acid protects our cells against damage from oxidation, one of the causes of aging. So natural selection probably selected for higher uric acid levels in our ancestors, even though some people end up getting gout, because those higher levels are especially useful in a species that lives as long as we do."

"So high levels of uric acid prevent aging?"

"Basically, that seems to be right. So far, however, there is no evidence that individuals with high uric acid levels live an especially long time, and in any case you don't want your toe to stay like that, so we are going to go ahead and get your uric acid levels down to the normal range to get the gout under control."

"Sounds sensible to me, Doc."

This is not an isolated example. A Darwinian perspective can already assist in the management of many medical conditions. Take strep throat:

"Well, it's strep all right, so you will need to take some penicillin for seven days," says the Darwinian physician.

"That will make me better faster, right?" the patient says hoarsely.

"Probably, and it will also make it less likely that you will develop diseases like rheumatic fever because of your body making immune substances that attack the bacteria."

"But why doesn't my body know better than to make substances that will attack my own heart?"

"Well, the streptococcus has evolved along with humans for millions of years, and its trick is to imitate the codes of human cells. So when we make antibodies that attack the strep bacteria, those antibodies are prone to attack our own tissues as well. We are in a contest with the strep organism, but we can't win because the strep evolves much faster than we do. It has a new generation every hour or so, while we take twenty years. Thank goodness we can still kill it with antibiotics, although this may be a temporary blessing. You will do yourself and the rest of the world a favor by taking your antibiotics even after you feel better, because otherwise you may be giving a lift to those variants that can survive short exposures to antibiotics, and those antibiotic-resistant organisms make life difficult for us all."

"Oh, now I see why I have to take the whole bottle. Okay."

Or take a patient who has had a heart attack:

"So, Doctor, if my high cholesterol is caused by my genes, what good will it do to change my diet?"

"Well, those genes aren't harmful in the normal environment we evolved in. If you spent six or eight hours walking around each day to find food, and if most of your food was complex starches and very lean meat from wild game, you wouldn't get heart disease."

"But how come I crave exactly the foods you say I shouldn't eat? No potato chips, no ice cream, no cheese, no steak? You medical types want to take away all the foods that taste best."

"I'm afraid we were wired to seek out certain things that were essential in small amounts but scarce on the African savannah. When our ancestors found a source of salt, sugar, or fat, it was usually a good idea for them to eat all they could get. Now that we can easily get any amount of salt, sugar, or fat just by tossing things into the grocery cart, most of us eat more than twice as much fat as our ancestors did, and lots more salt and sugar. You are right, it is a kind of a cruel joke—we do want exactly those things that are bad for us. Eating a healthy diet does not come naturally in the modern environment. We have to use our brains and our willpower to compensate for our primitive urges."

"Well, I still don't like giving up my favorite foods, but at least that makes it understandable."

There are a hundred more examples: advice given to a patient with a cold or diarrhea; an explanation of aging; the significance of morning sickness during pregnancy; the possible utility of allergy. While most medical conditions have yet to be explored from an evolutionary view, Darwinian medicine can already be useful in the clinic.

A caveat is necessary. Doctors and patients, like all other people, are prone to extend theories too far. We have lost count of how many reporters have called asking, "So you're saying we should not take aspirin for a fever, right?" Wrong! Clinical principles of medicine should come from clinical research, not from theory. It is a mistake to avoid aspirin just because we know that fever can be useful, and a mistake not to treat the unpleasant symptoms of some cases of pregnancy sickness, allergy, and anxiety. Each condition needs to be studied separately and each case considered individually. An evolutionary approach does, however, suggest that many such treatments are unnecessary or harmful and that we should do the research to see if the benefits are worth the costs.

PUBLIC POLICY IMPLICATIONS

We have said before, but here repeat, that moral principles cannot be deduced from biological facts. For instance, the knowledge that aging and death are inevitable has no direct implications for how much of our medical dollar we should spend on the very elderly. Facts can, however, help us to achieve whatever goals we decide to strive for. The current crisis in funding and organization of health care in the United States comes from several sources, including new funding mechanisms, new technology, other economic changes, and social values that increasingly condemn gross differences in the quality of health care. In a system this complex, no general policies will please everyone, and it may not be possible to implement the best available policies because of the power of politics.

While not pretending to offer solutions, we observe that the many participants in this debate don't even agree on what disease is. They know disease is bad but differ wildly on where it comes from and the extent to which it can be prevented or relieved. Some blame faulty genes, others emphasize the amount of disease that results from unfortunate human predilections, especially poor diets and drug use. According to one recent authoritative article, more than 70 percent of morbidity and mortality in the United States is preventable. The article argues strongly for investing in prevention because it will pay off in reduced health care costs. What a terrible irony and frightening harbinger it is that such a noble and practical proposal to improve human health has to be couched as a way to save money! In the light of history, however, this approach is understandable. Again and again, panels of distinguished physicians and researchers have called for prevention instead of treatment. The field of preventive medicine now provides some help, especially in matters of public policy, but people still do not get reliable advice from their physicians about how to stay healthy. New ways of organizing medical care may finally provide incentives for dedicating substantial clinical resources to preserving health based on principles of Darwinian medicine.

PERSONAL AND
PHILOSOPHICAL IMPLICATIONS

Few things are as important to us as our health. "How are you?" we ask in greeting each other, the convention of the inquiry still not completely covering its seriousness. "I've still got my health," says the person who has lost everything else. Health is vital. Without it, little else matters. We all want to understand the causes of disease to preserve and improve our health.

Long before there were effective treatments, physicians dispensed prognoses, hope, and, above all, meaning. When something terrible happens—and serious disease is always terrible—people want to know why. In a pantheistic world, the explanation was simple—one god had caused the problem, another could cure it. In the time since people have been trying to get along with only one God, explaining disease and evil has become more difficult. Generations of theologians have wrestled with the problem of theodicy—how can a good God allow such bad things to happen to good people?

Darwinian medicine can't offer a substitute for such explanations. It can't provide a universe in which events are part of a divine plan, much less one in which individual illness reflects individual sins. It can only show us why we are the way we are, why we are vulnerable to certain diseases. A Darwinian view of medicine simultaneously makes disease less and more meaningful. Diseases do not result from random or malevolent forces, they arise ultimately from past natural selection. Paradoxically, the same capacities that make us vulnerable to disease often confer benefits. The capacity for suffering is a useful defense. Autoimmune disease is a price of our remarkable ability to attack invaders. Cancer is the price of tissues that can repair themselves. Menopause may protect the interests of our genes in existing children. Even senescence and death are not random, but compromises struck by natural selection as it inexorably shaped our bodies to maximize the transmission of our genes. In such paradoxical benefits, some may find a gentle satisfaction, even a bit of meaning—at least the sort of meaning Dobzhansky recognized. After all, nothing in medicine makes sense except in the light of evolution.

Notes

Chapter 1. The Mystery of Disease

Page

6–7 For further discussion of proximate and ultimate (evolutionary) causation, see Ernst Mayr's *The Growth of Biological Thought* (Cambridge, Mass.: Belknap Press, 1982) or his brief article "How to Carry Out the Adaptationist Program," *American Naturalist*, 121:324–34 (1983). The problem of recognizing and confirming adaptations is dealt with on pp. 38–45 of George Williams' *Natural Selection* (New York: Oxford Univ. Press, 1992). A terminological revision is suggested by Paul W. Sherman in *Animal Behavior*, 36:616–19 (1988).

11 A history of social thought on Darwinism and of political uses of Darwinian metaphors is provided by Carl N. Degler's *In Search of Human Nature: The Decline and Revival of Darwinism in American Social Thought* (New York: Oxford Univ. Press, 1991). The inscription on the statue at Saranac Lake is quoted on page 410 of René Dubos's *Man Adapting* (New Haven: Yale Univ. Press, 1980).

Chapter 2. Evolution by Natural Selection

13 The Aristotle quotation is from p. 103 of *Aristotle: Parts of Animals*, translated by A. L. Peck (Cambridge, Mass.: Harvard Univ. Press, 1955).

Two recent books offer superb treatments of the modern concept of evolutionary adaptation. They are Helena Cronin's *The Ant and the Peacock* (New York: Cambridge Univ. Press, 1991) and Matt Ridley's *The Red Queen* (London, New York: Viking-Penguin, 1993). Cronin's account is more explicitly historical, with many quotations from Darwin, Wallace, and others. Both can be read with profit by both professional biologists and amateur naturalists.

13–14   The moth population that quickly evolved a darker color as its background darkened is discussed in many general works on evolution, for instance, on p. 58 of D. J. Futuyma's *Evolutionary Biology*, 2nd ed. (Sunderland, Mass.: Sinauer, 1986).

14   Examples of increased reproductive effort causing increased mortality or other costs are summarized on pages 28–9 and 188–93 of S. C. Stearns's *The Evolution of Life Histories* (New York: Oxford Univ. Press, 1992).

16–17   W. D. Hamilton's classic work is in *Journal of Theoretical Biology*, 7:1–52 (1964). Any modern book on evolution or animal behavior will discuss Hamilton's work. Richard Dawkins's book *The Selfish Gene*, new edition (Oxford: Oxford Univ. Press, 1989), offers a superb introduction to these ideas. The classic works on reciprocity are by R. L. Trivers in *Quarterly Review of Biology*, 46:35–57 (1971), and R. M. Axelrod's *The Evolution of Cooperation* (New York: Basic Books, 1984). These works are routinely reviewed in modern treatments of animal behavior, such as John Alcock's *Animal Behavior: An Evolutionary Approach*, 4th ed. (Sunderland, Mass.: Sinauer, 1989).

17   See E. O. Wilson's *Sociobiology* (Cambridge, Mass.: Harvard University Press, 1975) and *On Human Nature* (Cambridge, Mass.: Harvard Univ. Press, 1978) and Richard Alexander's *Darwinism and Human Affairs* (Seattle: University of Washington Press, 1979) and *The Biology of Moral Systems* (New York: Aldine de Gruyter, 1987).

17   The replay-the-tape-of-life idea is from pages 45–8 of S. J. Gould's *Wonderful Life: The Burgess Shale and the Nature of History* (New York: Norton, 1989).

18   The classic study of wing lengths of birds killed by a storm is cited in many recent works, such as John Maynard Smith's *Evolutionary Genetics* (New York: Oxford Univ. Press, 1989), which also explains the general topic of selection in favor of intermediate values (normalizing selection). For more on the optimization concept, see G. A. Parker and John Maynard Smith's article in *Nature*, 348:27–33 (1990), and *The Latest on the Best: Essays on Evolution and Optimality*, edited by John Dupré (Cambridge, Mass.: MIT Press, 1987).

21   The term *adaptationist program* was first used, disparagingly, by S. J. Gould and R. C. Lewontin in their much-cited article "The Spandrels of San Marco and the Panglossian Paradigm: A Critique of the Adaptationist Programme," *Proceedings of the Royal Society of London*, B205:581–98 (1979).

22   Gary Belovsky's work is described in *American Midland Naturalist*, 111:209–22 (1984).

22–23   For some clear thinking on the clutch-size problem and an introduction to recent work, see Jin Yoshimura and William Shield's article in *Bulletin of Mathematical Biology*, 54:445–64 (1992).

23 It must be that Darwin and his followers seldom found themselves at stag dances or singles bars, because the obvious minority-sex advantage somehow escaped their notice until it was pointed out by R. A. Fisher on p. 159 of his 1930 book *The Genetical Theory of Natural Selection* (New York: Dover, 1958 reprint).

24 For very recent work that takes an evolutionary view of disease see G. A. S. Harrison, ed. (1993), *Human Adaptation*, and *The Anthropology of Disease* by C. Mascie-Taylor (both Oxford: Oxford Univ. Press, 1994).

Chapter 3. Signs and Symptoms of Infectious Disease

27–29 The recent understanding of the role of fever in controlling infection is discussed in M. J. Kluger's article in *Fever: Basic Measurement and Management*, edited by P. A. MacKowiac (New York: Raven Press, 1990). For an older but still valuable overview, see his *Fever, Its Biology, Evolution, and Function* (Princeton, N.J.: Princeton Univ. Press, 1979). Data on acetaminophen's effects on chicken pox are presented by T. F. Doran and collaborators in an article in *Journal of Pediatrics*, 114:1045–8 (1989). The experiments on fever reduction and the progress of a cold are discussed by N. M. Graham and collaborators in *Journal of Infectious Disease*, 162:1277–82 (1990). The quotation on p. 28 is from Joan Stephenson in *Family Practice News*, 23:1, 16 (1993).

29–31 The sequestration of iron as a defense against bacterial pathogens is discussed by E. D. Weinberg in *Physiological Reviews*, 64:65–102 (1984). The treatment of malaria with iron chelating agents is reported by V. Gordeuk et al. in *The New England Journal of Medicine*, 327:1473–7 (1992).

31–33 For a wide-ranging review of progress in bringing evolution to bear on microbiology, see *Parasite-Host Associations: Coexistence or Conflict*, edited by C. A. Toft et al. (New York: Oxford Univ. Press, 1991). A still-valuable general review of host-parasite coevolution is P. W. Price's *Evolutionary Biology of Parasites* (Princeton, N.J.: Princeton Univ. Press, 1980).

33–36 Behavioral defenses against parasites are discussed by B. L. Hart in *Neuroscience and Biobehavioral Reviews*, 14:273–94 (1990). The functions of pain and the shortened lives of those who lack it are described by Ronald Melzack in *The Puzzle of Pain* (New York: Basic Books, 1973).

36 The biocidal action of tears is discussed by S. Hassoun in *Allergie et Immunologie*, 25:98–100 (1993), and that of saliva by D. J. Smith and M. A. Taubman in *Critical Reviews of Oral Biology and Medicine*, 4:335–41 (1993).

36   A relevant article on nasal sprays is provided by R. Dockhorn and collaborators in *Journal of Allergy and Clinical Immunology*, 90:1076–82 (1992).

37   Important works on the psychology of food aversions and related defenses are provided by M. E. P. Seligman in *Psychological Review*, 77:406–18 (1970), and by John Garcia and F. R. Ervin in *Communications in Behavioral Biology*, (A)1:389–415 (1968).

37–38   The diarrhea article is by H. L. DuPont and R. B. Hornick in *Journal of the American Medical Association*, 226:1525–8 (1973).

38–39   Profet's theory is presented in *Quarterly Review of Biology*, 68:335–86 (1993). Strassman's paper was presented to the 1994 Human Behavior and Evolution Society.

39–40   A good general introduction to immunology is Chapter 16 of *Life: The Science of Biology*, 3rd ed., by W. K. Purves, G. H. Orians, and H. C. Heller (Sunderland, Mass.: Sinauer, 1992).

41   Many dramatic examples of the ravages of parasitic diseases are described, and some pictured, by Michael Katz et al. in *Parasitic Diseases*, 2nd ed. (New York: Springer, 1989).

42   Hemoglobin increase in compensation for decreased lung function is described on pages 307 and 418 of A. J. Vander et al.'s *Human Physiology: The Mechanisms of Body Function*, 5th ed. (New York: McGraw-Hill, 1990).

42–44   For a readable and authoritative introduction to deceptive strategies used by pathogens, see Ursula W. Goodenough's article in *American Scientist*, 79:344–55 (1991). Specifically antimalarial strategies are discussed by D. J. Roberts et al. in *Nature*, 357:689–92 (1992). A wealth of material on autoimmune disease is provided by *The Autoimmune Diseases*, vol. 2, edited by N. R. Rose and I. R. Mackay (San Diego: Academic, 1992). See especially the introductory chapter by Rose and Mackay. The relationship between obsessive-compulsive disorder and Sydenham's chorea is discussed by Judith Rapaport on pages 83–89 of *Scientific American* (March 1989).

44–45   Reactions and overreactions to a bacterial toxin are discussed by E. K. LeGrand in *Journal of the American Veterinary Medical Association*, 197:454–6 (1990).

45   The best evolutionary perspective on AIDS is in P. W. Ewald's *Evolution of Infectious Disease* (New York: Oxford Univ. Press, 1993). See also B. R. Levins's article on pp. 101–11 of *AIDS, the Modern Plague*, edited by P. A. Distler et al. (Blacksburg, Va.: Presidential Symposium, Virginia Polytechnic Institute and State University, 1993).

45–47   Viral alteration of host cell structure is discussed by Shmuel Wolf et al. in *Science*, 246:377–9 (1989). Fungal castration of plants is reviewed by Keith Clay in *Trends in Ecology and Evolution*, 6:162–6 (1991). Behavior manipulation by the rabies virus is discussed by G. M. Baer in *The Natural History of Rabies* (New York: Academic, 1973). A general review of manipulation of host behavior by parasites is provided by A. P. Dobson in *The Quarterly Review of Biology*, 63:139–65 (1988). Many medically important examples of host manipulation are discussed by Heven in *The Host-Invader Interplay*, edited by H. Van den Bossche (Amsterdam: Elsevier/North Holland, 1980).

47–48   Ewald's article, mentioned in our Preface, is "Evolutionary Biology and the Treatment of Signs and Symptoms of Infectious Disease," *Journal of Theoretical Biology*, 86:169–76. It forms the basis for Table 3-1. Professional conferences on evolutionary approaches to medicine include one in Boston at the February 1993 meeting of the American Association for the Advancement of Science, another at the London School of Economics in June 1993.

Chapter 4. An Arms Race Without End

49   The classic work on biological arms races is Richard Dawkins and J. L. Krebbs' article in *Proceedings of the Royal Society of London*, B105:489–511. Alice's race with the Red Queen is in Chapter 2 of *Through the Looking Glass* by Lewis Carroll.

50   The account of President Coolidge's son's death and its emotional and political effects is taken from p. 14 of an article by R. S. Robins and M. Dorn in *Politics and the Life Sciences*, 12:3–17 (1993).

50   A superb popular account of the nature and power of natural selection is Richard Dawkins' *The Blind Watchmaker* (New York: W. W. Norton, 1986).

52   Evidence of devastation of native populations by introduced diseases is summarized by R. M. Anderson and R. M. May's *Infectious Diseases of Humans* (New York: Oxford Univ. Press, 1991) and by F. L. Black in *Science*, 258:1739–40 (1992).

53–56   The quotation is from M. L. Cohen's article in *Science*, 257:1050–5 (1992). Useful recent reviews of bacterial resistance to antibiotics are provided by J. P. W. Young and B. R. Levin in their article in *Genes in Ecology*, edited by R. J. Berry et al. (Boston: Blackwell Scientific, 1991) and by S. B. Levy's *The Antibiotic Paradox: How Miracle Drugs Are Destroying the Miracle* (New York: Plenum, 1992). See also Rick Weiss in *Science*, 255:148–50. The use of antibiotics in livestock is discussed by S. B. Levy in *The New England Journal of Medicine*, 323:335–37, 1990. Our data on tuberculosis are mainly from B. R. Bloom and C. J. L. Murray in *Science*, 257:1055–64. The 1969 quote from the Surgeon General is in Bloom's article. H. C. Neu's work is in *Sci-*

ence, 257:1064–73 (1992). The article by Ridley and Low is in *The Atlantic, 272*(3):76–86 (September 1993).

57–58 Three examples of authoritative statements on an inevitable evolutionary reduction of virulence provide epigraphs for the first chapter of Paul W. Ewald's book cited for p. 45. One not cited by Ewald is the distinguished population geneticist Theodosius Dobzhansky's assertion that parasitism "is a form of relationship which is unstable in the evolutionary sense, and . . . it will tend to disappear and be replaced by cooperation and mutualism," *Genetics and the Origin of Species*, 3rd ed. (New York: Columbia Univ. Press, 1951, p. 285). Genetic diversity of HIV within a host is documented by several writers in *Science, 254*:941, 963–9 (1991); *255*:1134–7 (1992). Genetic diversity of a parasitic helminth population in a single host is documented by M. Mulvey et al. in *Evolution, 45*:1628–40 (1991). The data on fluke infections of fig wasps are in E. A. Herre's article in *Science, 259*:1442–5 (1993).

58–60 There is a large literature on the different effects of selection within and between populations. The special case of selection on parasites within and between hosts is modeled by R. L. Anderson and R. L. May's book cited for p. 52. J. J. Bull and I. J. Molineux's experimental verification of the expected increase in virulence of a virus that had its fitness decoupled from that of its host is presented in *Evolution, 46*:882–95 (1992). Other important works are R. B. Johnson's in *Journal of Theoretical Biology, 122*:19–24 (1986), and S. A. Frank's in *Proceedings of the Royal Society of London, B259*:195–7 (1992).

60 Our favorite account of the Semmelweis story is the 1909 classic by William J. Sinclair, *Semmelweis, His Life and His Doctrine* (Manchester: The University Press).

62–63 A good introduction to mimicry is provided by J. R. G. Turner's article on pp. 141–61 of *The Biology of Butterflies*, edited by R. I. Vane-Wright and P. R. Ackery (London and Orlando: Academic, 1984). Works on molecular mimicry and related phenomena are cited for pp 43–44.

63–64 Most of our information on the effects of novel environments on infection is from R. M. Krause's article in *Science, 257*:1073–8 (1992). Detailed data on the Ebola virus are provided by P. H. Sureau's article in *Reviews of Infectious Diseases, 11*(4):790s–793s (1989).

Chapter 5. Injury

66 The quotation at the beginning of the chapter is from Chapter 6 of Mark Twain's *The Adventures of Huckleberry Finn*.

68 John Garcia's classic work, with F. R. Ervin, is cited for p. 37.

68–69   The work on monkeys' conditioned fear of snakes is by Susan Mineka and collaborators in *Animal Learning and Behavior*, 8:653–63 (1980).

69–70   Repair of mechanical damage is discussed by P. L. McNeil in *American Scientist*, 79:222–35, and by Natalie Angier in *The New York Times*, November 9, 1993, pp. C1, C14.

70-71   Many of the special aspects of burn healing are discussed in *Burn Care and Rehabilitation: Principles and Practice*, edited by R. L. Richard and M. J. Staley (Philadelphia: F. A. Davis, 1994). See especially Chapter 5, by D. G. Greenhalgh and M. J. Staley.

72   The trout-hatchery story and a general discussion of damage by sunlight are provided by Alfred Perlmutter in *Science*, 133:1081–2 (1961).

73–75   UV-B effects on Langerhans cells are discussed by M. Vermeer et al. in *Journal of Investigative Dermatology*, 97:729–34 (1991). An epidemiological study of the increase in melanoma rates is provided by J. M. Elwood and collaborators in *International Journal of Epidemiology*, 19:801–10 (1990). A less technical discussion, with emphasis on immunological aspects of melanoma, is David Concal's in *New Scientist*, 134:23–8 (1991). Interactions between the nervous system and Langerhans cells are discussed by J. Hosoi et al. in *Nature*, 159–63 (1993). The role of sunscreens in causing excess exposure to UV-A is discussed by P. M. Farr and B. L. Diffey in *The Lancet*, 1(8635):429–31 (1989). Eye damage by sunlight is discussed by L. Semes in *Optometry Clinics*, 1(2):28–34 (1991). The beneficial effects of sunscreen use are reported by S. C. Thompson and collaborators in *The New England Journal of Medicine*, 329:1147–51 (1993).

75–76   The work of R. J. Goss in the *Journal of Theoretical Biology*, 159: 241–60 is a good introduction to the literature and current controversies on the evolution of regeneration.

Chapter 6. Toxins: New, Old, and Everywhere

77   Works by McNeil and by Angier, cited for pp. 69–70, are relevant to the sort of damage the whisky caused to Don Birnham's stomach.

78–81   An introduction to the work of Bruce Ames et al. is provided in a 1991 response by Ames and L. S. Gold to criticisms of their earlier work (*Science*, 251:607–8). Timothy Johns' *With Bitter Herbs They Shall Eat It* (Tucson: Univ. of Arizona Press, 1990), reviews many aspects of human ecology in relation to plant toxins. It also details a fascinating history of human dealings with potatoes and their toxins, and of medicinal uses of plant toxins. A more technical work is *Toxic Plants*, edited by A. D. Kinghorn (New York:

Columbia Univ. Press, 1979). An early but unexcelled review of chemical defenses in arthropods is by Thomas Eisner on pp. 157–217 of *Chemical Ecology*, edited by Ernest Sondheimer and J. B. Simeone (New York: Academic, 1970). The first serious discussion of trade-offs between chemical defenses and other values, such as rapidity of development, was by G. H. Orians and D. H. Janzen in the *American Naturalist*, 108:581–92 (1974). For a dramatic account of plant defenses, with details of electrical signaling and rapid adaptation, see Paul Simons' *The Action Plant* (Boston: Blackwell, 1992). It includes a discussion of the role of aspirinlike hormones in plants.

80 Our interpretation of nectar toxins is based on D. F. Rhoades and J. C. Bergdahl's article in *American Naturalist*, 117:798–803 (1981).

81 A dramatic account of the consequences of fungal toxins for human life is provided by Mary K. Matossian's *Poisons of the Past: Molds, Epidemics, and History* (New Haven: Yale Univ. Press, 1989).

82–83 The high incidence of PTC tasting in the Peruvian Andes is documented by R. M. Barruto and coauthors in *Human Biology*, 47:193–9 (1975). The study of oxylate kidney stones is that of G. C. Curhan et al. in *The New England Journal of Medicine*, 328:833–8 (1993). Our kidney-stone discussion is also based on that of S. B. Eaton and D. A. Nelson, "Calcium in Evolutionary Perspective," in *American Journal of Clinical Nutrition*, 54:281s–287s. For a wide-ranging review of the evolution of chemical and other defense mechanisms, see D. H. Janzen's article on pages 145–64 of *Physiological Ecology: An Evolutionary Approach to Resource Use*, edited by C. R. Townsend and Peter Calow (Oxford: Blackwell, 1981).

84 Maize processing is described by S. H. Katz et al., in *Science*, 184:765–73 (1973).

84 The information on tannins in acorns and the detoxification of arum by cooking is from pp. 63–5 in Timothy Johns's book cited for pp. 78–80.

85 The toxicity of disease-resistant potatoes is discussed on pages 106–59 of Johns's book cited for pages 78–80.

86 Bacterial resistance to antibiotics in people with dental fillings is discussed by A. O. Summers et al. in *Antimicrobial Agents and Chemotherapy*, 37:825–34 (1993). Examples of unrealistic arguments on environmental toxins can be found in *Biosphere Politics* (New York: Crown, 1991) and other works by Jeremy Rifkin.

87–89 The antiteratogen theory of morning sickness is presented by Margie Profet on pp. 327–65 of *The Adapted Mind*, edited by J. H. Barkow et al. (New York: Oxford Univ. Press, 1992).

89   The reluctance of regulatory agencies to take fetal sensitivities into account is discussed by Ann Gibbons in *Science*, 254:25 (1991).

Chapter 7. *Genes and Disease: Defects, Quirks, and Compromises*

92–94   A recent general treatment of medical genetics is T. D. Gelehrter and F. S. Collins' *Principles of Medical Genetics* (Baltimore: Williams & Wilkins, 1990). A number of articles describing advances in the understanding of genetic diseases and progress in gene therapy were published in 1992 and 1993 in *Science* (256:773–813, 258:744–5, 260:926–32). For a vivid personal view of the development of modern medical genetics and wise commentary on its implications, we recommend James Neel's *Physician to the Genome* (New York: Wiley, 1994). Another thoughtful treatment of the ethics of genetic counseling can be found in *Genetic Disorders and the Fetus*, edited by Aubrey Milunsky (Baltimore: Johns Hopkins Univ. Press, 1992); see especially the chapter by J. C. Fletcher and D. C. Wertz.

96–99   Selection against unfavorable genes, their rate of loss by such selection, their expected equilibrium frequencies in populations, and related quantities can be related to one another algebraically, as explained by any textbook of population genetics, such as J. Maynard Smith's *Evolutionary Genetics* (New York: Oxford Univ. Press, 1989). Our presentation in this chapter is greatly simplified. *Huntington's Disease*, edited by P. S. Harper (London: Saunders, 1991), summarizes the history and epidemiology of this condition. It would be difficult to find a modern textbook of genetics or evolution that does not discuss the sickle-cell gene. Our favorite treatment is by Jared Diamond in *Natural History*, June 1988, pp. 10–13.

100–101   Our information on G6PD deficiency is from an article by Ernest Beutler in *The New England Journal of Medicine*, 324:169–74 (1991). The quotation from F. S. Collins is from his article in *Science*, 774 (1992). Complications in cystic fibrosis genetics are reviewed by Gina Kolata in *The New York Times*, November 16, 1993, pp. C1, C3, and related evolutionary problems by Natalie Angier in *The New York Times*, June 1, 1994, p. B9. Contributions to the study of Tay-Sachs disease are offered by B. Spyropoulos and Jared Diamond in *Nature*, 331:666 (1989); by S. J. O'Brien in *Current Biology*, 1:209–11 (1991); and by N. C. Myrianthopoulos and Michael Melnick in "Tay-Sachs Disease: Screening and Prevention," in *Palm Springs International Conference on Tay-Sachs Disease* edited by M. M. Kaback (New York: Liss, 1977). Our information on the human fragile-X syndrome is from F. Vogel et al.'s article in *Human Genetics*,

86:25–32 (1990). Jared Diamond has written a number of nicely reasoned articles on hidden benefits of genes that cause disease. Some of these are in *Discover*, November 1989, pp. 72–8, and in *Natural History*, June 1988, pp. 10–13, and February 1990, pp. 26–30. Worthy examples of the voluminous literature on the genetic aspects of disease and health are Teresa Costa et al.'s article in *American Journal of Human Genetics*, 21:321–42 (1985), and in a group of five articles on anthropological aspects of genetic disease in *American Journal of Physical Anthropology*, 62(1) (1983).

101   The effect of PKU on miscarriage rates is discussed by L. I. Woolf et al. in *Annals of Human Genetics*, 38:461–9 (1975). A recent statement of Richard Dawkins' idea that a body is the genes' way of making more genes is his *The Selfish Gene*, new ed. (New York: Oxford Univ. Press, 1989).

101–102   The fitness effects of the T-locus in mice are discussed by Patricia Franks and Sarah Lenington in *Behavioral Ecology and Sociobiology*, 18:395–404 (1986). Medical aspects of mitochondrial DNA are discussed by Angus Clarke in *Journal of Medical Genetics*, 27:451–6 (1990). For general treatments of intragenomic conflict, see Leda Cosmides et al.'s classic work in *Journal of Theoretical Biology*, 89:83–129 (1981), and David Haig and Alan Grafen's article in *Journal of Theoretical Biology*, 153:531–58 (1991).

102–103   Familial and environmental aspects of cardiac malfunction are discussed by M. P. Stern on pp. 93–104 in *Genetic Epidemiology of Coronary Heart Disease: Past, Present, and Future*, edited by M. P. Stern (New York: Liss, 1984).

103–105   Piggy's extreme dependence on his glasses, and the tragic results of their damage and spiteful theft, are depicted in Chapters 10 and 11 of *Lord of the Flies* by William Golding. The quotation is from Chapter 11. The sudden emergence of myopia in the children of urbanized Eskimos is documented by F. A. Young et al. in *American Journal of Ophthalmology*, 46:676–85 (1969). General discussions of the genetics and etiology of myopia are provided by Elio Raviola and T. N. Wiesel's article in *The New England Journal of Medicine*, 312:1609–15 (1985); by B. J. Curtin's *The Myopias* (Philadelphia: Harper & Row, 1988); and by G. R. Bock and Kate Widdows in *Myopia and the Control of Eye Growth* (Chichester, New York: Wiley, 1990). A brief summary of recent research is provided by Jane E. Brody in *The New York Times*, June 1, 1994, p. C10.

105   Information on the genetics of alcoholism is in M. A. Schickit's article in *Journal of the American Medical Association*, (1985); in J. S. Searles' in *Journal of Abnormal Psychology*, 97:153–67 (1988); and in M. Mullen's in *British Journal of Addictions*, 84:1433–40 (1989).

106 The quotations are from pp. 89–90 of Melvin Konner's *The Tangled Wing: Biological Constraints on the Human Spirit* (New York: Harper Colophon, 1983) and p. 215 of Richard Dawkins' *The Selfish Gene* (New York: Oxford Univ. Press, 1976).

Chapter 8. Aging as the Fountain of Youth

107 The Irish ballad is on p. 103 of *100 Irish Ballads* (Dublin: Walton's, 1985). For the general reader, an excellent overview of the evolution of aging is provided by several articles in the February 1992 issue of *Natural History* and by R. Sapolsky and Caleb Finch on pp. 30–8 of the March–April 1991 issue of *The Sciences*. Excellent recent technical works are available in M. R. Rose's article in *Theoretical Population Biology*, 28:342–58 (1984); in his *Evolutionary Biology of Aging* (New York: Oxford Univ. Press, 1991); and in Caleb Finch's *Longevity, Senescence, and the Genome* (Chicago: Univ. of Chicago Press, 1991).

108–109 Death rates in the United States are from *Vital Statistics in the United States, 1989* (Washington, D.C.: U.S. National Center for Health Statistics, 1992). The demographic aspects of aging are well reviewed by J. F. Fries and L. M. Crapo in *Vitality and Aging* (San Francisco: Freeman, 1981).

110 Figure 8-1 is redrawn from Figure 3-2 in *Vitality and Aging* with permission.

111–112 Figure 8-3 is redrawn from Figure 9.2 in *Vitality and Aging* with permission. We got the story about people fleeing a tiger from Helena Cronin's *The Ant and the Peacock* (New York: Cambridge Univ. Press, 1992).

111–112 The lines about the "one-hoss shay" are from "The Deacon's Masterpiece" on pp. 158–60 of *The Complete Poetical Works of Oliver Wendell Holmes* (Boston: Houghton Mifflin, 1908). The apparent coordination of aging effects is discussed by B. L. Strehler and A. S. Mildvan in *Science*, 132:14–21 (1960).

113 The quotation is from August Weismann's "The Duration of Life," in *A. Weismann: Essays upon Heredity and Kindred Biological Problems*, edited by E. B. Poulton et al. (Oxford: Clarendon Press, 1891–2). The article by G. C. Williams is in *Evolution*, 11:398–411 (1957).

113–114 The J. B. S. Haldane reference is to *New Paths in Genetics* (New York: Harper, 1942). The P. B. Medawar quotation is from p. 38 of his article "Old Age and Natural Death," reprinted on pp. 17–43 of his *The Uniqueness of the Individual* (London: Methuen, 1957). See also his *An Unsolved Problem in Biology* (London: M. K. Lewis, 1952). The classic theoretical treatment of the subject is W. D. Hamilton's in *Journal of Theoretical Biology*, 12:12–45 (1968).

114–115 For important recent comments on the evolution of menopause, see A. R. Rogers' article in *Evolutionary Ecology*, 7:406–20, Kim Hill and A. M. Hurtado's in *Human Nature*, 2:313–50 (1991), S. N. Austad in *Experimental Gerontology*, 29:255–63 (1994). Alex Comfort's book is *The Biology of Senescence*, 3rd ed. (New York: Elsevier, 1979).

115–116 Figure 8-4 is adapted from R. M. Nesse's article in *Experimental Gerontology*, 23:445–53 (1988). R. L. Albin's article is in *Ethology and Sociobiology*, 9:371–82 (1988). Hemochromatosis is reviewed by J. F. Desforges in the *New England Journal of Medicine*, 328:1616–20 (1993).

116–117 For recent findings on the genetics of Alzheimer's disease, see the article by W. Strittmatter et al. in *Proceedings of the National Academy of Sciences* (*U.S.*), 90:1977–81 (1993). S. I. Rapoport's work is in *Medical Hypotheses*, 29:147–50.

117 R. R. Sokal's and other experimental studies of the role of pleiotropic genes in senescence are summarized in M. R. Rose's book, cited for the beginning of the chapter. See especially his pp. 50–6 and 179–80.

118–120 Work on dietary restriction is reviewed by J. P. Phelan and S. N. Austad in *Growth, Development, and Aging*, 53(1–2):4–6 (1989). For evidence on the beneficial effects of antioxidants and their mechanism of action, see R. G. Cutler's article in *American Journal of Clinical Nutrition*, 53:373s–379s (1991). The quotation on gout is from p. 622 of Lubert Stryer's *Biochemistry*, 3rd ed. (New York: Freeman, 1988). S. N. Austad's reasons for believing that the aging process may be quite different in different species are presented in *Aging*, 5:259–67 (1994). His opossum work is in *Journal of Zoology*, 229:695–708 (1994).

122 E. T. Whittaker's discussion of postulates of impotence is mainly on pp. 58–60 of his *From Euclid to Eddington. A Study of Conceptions of the External World* (New York: Dover, 1958).

Chapter 9. Legacies of Evolutionary History

For authoritative and accessible overviews of human evolution, we suggest Roger Lewin's *In the Age of Mankind: A Smithsonian Book of Human Evolution* (Washington, D.C.: Smithsonian Books, 1988) and Jared Diamond's *The Third Chimpanzee* (New York: HarperCollins, 1992). For an engrossing biography of a contemporary hunter-gatherer woman, we recommend Marjorie Shostack's *Nisa: The Life and Words of a !Kung Woman* (New York: Vantage Books, 1983).

125 The quotation from Charles Darwin is from p. 191 of the first edition of *The Origin of Species* (London: John Murray, 1859).

125–127 A more dramatic account of the unfortunate effect of human speech adaptations on traffic control in the throat is provided in Chapter 10 of Elaine Morgan's *The Scars of Evolution* (London: Penguin, 1990). More technically detailed information can be found in Philip Lieberman and Sheila E. Blumstein's *Speech Physiology, Speech Perception, and Acoustic Phonetics* (Cambridge, England: Cambridge Univ. Press, 1988).

131 Our use of the book by George Estabrooks, *Man, The Mechanical Misfit* (New York: Macmillan, 1941), is at variance with its spirit. While it does describe many design flaws of the human body, its main message is the misfit between that design and the uses to which it is put in modern times. It is also a eugenicist tract.

135 "Stone Agers in the Fast Lane" is the title of an article by S. B. Eaton et al. in *The American Journal of Medicine*, 84:739–49 (1988).

136–137 Luigi Cavalli-Sforza et al. in *Science*, 259:639–46 (1993), estimate the current population at about a thousand times that of the Stone Age. The prevalence of human infanticide, and comparable behavior in other species, has recently gotten detailed attention. See *Infanticide: Comparative and Evolutionary Perspectives*, edited by G. Hausfater and S. B. Hrdy (New York: Aldine, 1984).

137 For details of the symptoms of protozoan and helminth diseases, see Part XV (pp. 1714–78) of *The Cecil Textbook of Medicine*, edited by J. B. Wyngaarden and L. H. Smith (Philadelphia: Saunders, 1982). Many of the unpleasant effects of parasites are described, and some pictured, in the book by M. Katz et al. cited for p. 41. Richard Alexander's quote is from p. 138 of the book cited for p. 17.

140 A 15,000-year antiquity for domesticated dogs is suggested by Vitaly Shevoroshkin and John Woodward in their article on pp. 173–97 in *Ways of Knowing. The Reality Club 3*, edited by John Brockman (New York: Prentice Hall, 1991).

142 The quotation about cave paintings is from p. 57 in Melvin Konnor's *The Tangled Wing: Biological Constraints on the Human Spirit* (New York: Harper Colophon, 1983).

Chapter 10. Diseases of Civilization

144–145 For more on the origins of agriculture and husbandry, see Chapters 10 and 14 of Jared Diamond's book, cited for the beginning of Chapter 9.

145–146 Use of wild plant products to cure scurvy is discussed by Ingolfur Davidsson in *Natturufraedingurinn*, 42:140–4 (1972). Nutritional deficiencies and other problems evident in the 1500-year-old Amerind skeletons are documented by J. Lallo et al. on pp. 213–38 of *Early Native Americans*, edited by D. L. Browman (The Hague and New York: Moulton, 1980).

147 The supernormal stimulus idea is discussed in many general works and textbooks, for instance, on pp. 27–9 of John Alcock's book cited for pp. 16–17.

148–149 For discussions of the role of dietary fat in modern medical problems, see H. B. Eaton's article in *Lipids*, 27:814–20 (1992); *Western Diseases, Their Emergence and Prevention*, edited by H. C. Trowell and D. P. Burkitt (Cambridge, Mass.: Harvard Univ. Press, 1981), and H. B. Eaton et al.'s *The Paleolithic Prescription* (New York: Harper and Row, 1988). For a convincing work on the profound role of environment in public health and the relative unimportance of medicine, see Thomas McKeown's *The Role of Medicine: Dream, Mirage, or Nemesis?* (Princeton, N.J.: Princeton Univ. Press, 1979).

149–150 The discussion of thrifty genotypes follows J. V. Neel's article in *Sorono Symposium*, 47:281–93 (1982), and Gary Dowse and Paul Zimmet's in *British Medical Journal*, 306:532–3 (1993). The effects of intermittent dieting are discussed in an article by J. O. Hill et al. in *International Journal of Obesity*, 12:547–55 (1988). The findings on artificial sweeteners are presented by D. Stellman and L. Garfinkel in *Preventive Medicine*, 15:195–202 (1986). Evidence for a long-term metabolic effect of intermittent food restriction is presented by G. L. Blackburn et al. in *American Journal of Clinical Nutrition*, 49:1105–9 (1989). Our conclusions and recommendations on diet and weight control summarize a detailed discussion published in a series of articles in *The New York Times*, November 22–5, 1992.

150 The incidence of dental caries in prehistoric Georgia is discussed by C. S. Larsen et al. in *Advances in Dental Anthropology*, edited by M. A. Kelley and C. S. Larsen (New York: Wiley-Liss, 1991).

151 For an example of a tribal society's use of a psychotropic drug, see Napoleon Chagnon's discussion of the use of *ebene* in Venezuela in *Yanomamo: The Last Days of Eden* (New York: Harcourt Brace Jovanovich, 1992).

152 The inheritance of susceptibility to substance abuse is discussed by C. R. Cloninger in *Archives of General Psychiatry*, 38:961–8 (1981); by M. A. Schuckit in *Journal of the American Medical Association*, 254:2614–7 (1985); and by J. S. Searles in *Journal of Abnormal Psychiatry*, 97:153–7 (1988). See also R. M. Nesse's article in *Ethology and Sociobiology* (in press).

154 Alan Weder and Nickolas Schork have published their theory in *Hypertension*, 24:145–56 (1994).

155–156 Skin color in relation to rickets is discussed by W. M. S. Russell in *Ecology of Disease*, 2:95–106 (1983). The rapid evolutionary loss of pigment and eyes by animals living in caves is discussed by R. W. Mitchell and collaborators in "Mexican Eyeless Fishes, Genus *Astyanax*: Environment, Distribution and Evolution,"

Special Publications. The Museum. Texas Tech University, 12:1–89 (1977). Evidence for the importance of introduced diseases in the destruction of New World peoples is summarized by F. L. Black in *Science*, 258:1739–40. See also the work of M. Anderson and R. M. May cited for p. 52.

Chapter 11. Allergy

A good introduction to pollen allergies is N. Mygind's *Essential Allergy* (Oxford: Blackwell, 1986). A more detailed review is in *Allergic Diseases: Diagnosis and Management*, edited by R. Patterson (Philadelphia: J. B. Lippincott, 1993). A useful book on pollen is R. B. Knox's *Pollen and Allergy* (Baltimore: University Park Press, 1978).

159   For details on the IgE system, see O. L. Frick's article on pp. 197–227 of *Basic and Clinical Immunology*, 6th ed., edited by D. P. Stites, J. D. Stobo, and J. V. Wells (Norwich, Conn.: Appleby and Lange, 1987), and C. R. Zeiss and J. J. Prusansky's on pp. 33–46 of *Allergic Diseases: Diagnosis and Management* (Philadelphia: J. B. Lippincott, 1993). Amos Bouskila and D. T. Blumstein provide a detailed discussion of what we call *the smoke-detector principle* in *American Naturalist*, 139:161–76 (1992).

160   The *New York Times* quotation is from section 6, p. 52, March 28, 1993. The textbook quoted is E. S. Golub's *Immunology: A Synthesis* (Sunderland, Mass.: Sinauer, 1987).

160–161   The history of ideas on the function of the ampullae of Lorenzini is discussed in a delightful article, "The Sense of Discovery and Vice Versa," by K. S. Thomson in *American Scientist*, 71:522–5 (1983). More recent work is reviewed by H. Wissing et al. in *Progress in Brain Research*, 74:99–107 (1988).

162–163   For discussions of IgE in relation to helminth infections, see A. Capron and J.-P. Dessaint's work in *Chemical Immunology*, 49:236–44 (1990), and K. Q. Nguyen and O. G. Rodman's in *International Journal of Dermatology*, 32:291–7 (1984).

163   Profet's article is in *Quarterly Review of Biology*, 66:23–62 (1991).

164–166   For more information on the apparently increasing incidence of allergy, see works by L. Gamlin in the June 1990 issue of *New Scientist* and by Ronald Finn in *Lancet*, 340:1453–5 (1992). The genetics of atopy is reviewed by J. M. Hopkins in *Journal of the Royal College of Physicians* (London), 24:159–60 (1990). Evidence of the prevalence of genetic deficiencies in detoxification enzymes is reviewed by M. F. W. Festing in *Critical Reviews in Toxicology*, 18:1–26. Unfortunately, most of the research relates to variation in detoxification of drugs, not to routinely encountered toxins.

169 The study of prevention of allergy is by S. H. Arshad et al. and is published in *Lancet*, 339:1493–97 (1992).

169–170 See citations for pp. 162–64 for indications of the increasing frequency of allergies. The redundancy and complexity of the immune system are well described in S. Ohno in *Chemical Immunology*, 49:21–34 (1990).

Chapter 12. Cancer

172–174 Our perspective on cancer derives from Leo Buss's book *The Evolution of Individuality* (Princeton, N.J.: Princeton Univ. Press, 1987). Liles's article is in *MBL Science*, 3:9–13 (1988).

175–177 Our account of the cellular, hormonal, and immunological mechanisms of cancer control is a greatly simplified retelling of that provided by two collections of articles in *Science*, 254:1131–73 (1991) and 259:616–38 (1993). The data on the p53 gene are from Elizabeth Culotta and D. E. Koshland's article in *Science*, 262:1958–61 (1993). Many of our statements on genetic factors in cancer are supported by Chapter 5 of D. M. Prescott and A. S. Flexner's *Cancer. The Misguided Cell*, 2nd ed. (Sunderland, Mass.: Sinauer, 1986). Cosmides and Tooby's observations were made in a talk presented to the 1994 meeting of the Human Behavior and Evolution Society.

178 On sunshine as carcinogen and its effects on the immune system, we recommend David Concar's easily readable account in the *New Scientist*, 134 (1821):23–8 (1992).

179–181 Our discussion of women's reproductive cancers summarizes that of W. B. Eaton et al. in *Quarterly Review of Biology*, 69:353–67 (1994). The reduction in uterine and ovarian cancer risk as a result of oral contraceptive use is documented by B. E. Henderson et al. in *Science*, 259:633–8 (1993).

Chapter 13. Sex and Reproduction

183–184 The current debate over the evolutionary origins of sex is well presented in Matt Ridley's *The Red Queen* (New York: Macmillan, 1993). For a more technical discussion, see R. E. Michod and B. R. Levin, editors, *The Evolution of Sex* (Sunderland, Mass.: Sinauer, 1988). For the parasite theory of sexuality, see W. D. Hamilton, R. Axelrod, and R. Tanese's article in *Proceedings of the National Academy of Sciences*, 87:3566–73 (1990). For some origins of the current debate, see G. C. Williams' *Sex and Evolution* (Princeton, N.J.: Princeton Univ. Press, 1975) and J. Maynard Smith's *The Evolution of Sex* (New York: Cambridge Univ.

Press, 1978). A recent review by S. Sarkar appears in *BioScience*, 42(6):448–54 (1992). The evolution of genetic diversity is reviewed by Wayne K. Potts and Edward K. Wakeland in *Trends in Ecology and Evolution*, 5:181–7 (1990)

184–185   For a discussion of why there are large eggs and small sperm, see pp. 151–5 of Maynard Smith's *The Evolution of Sex*, cited above. Pp. 130–9 of the same work present the currently accepted view of why some organisms are hermaphrodites and others have separate sexes. A more detailed treatment is found in E. L. Charnov's *The Theory of Sex Allocation* (Princeton N.J.: Princeton Univ. Press, 1982).

184–187   Current controversies on the theory of sexual selection, which deals with male-female differences in reproductive adaptations, are discussed in *Sexual Selection: Testing the Alternatives*, edited by J. W. Bradbury and M. B. Anderson (New York: Wiley-Interscience, 1987). The historical development and current form of this theory are beautifully presented by Helena Cronin's *The Ant and the Peacock* (New York: Cambridge Univ. Press, 1991).

187   The problems expected as a result of a female-biased sex ratio are discussed by P. Secord in *Personality and Social Psychology Bulletin*, 9(4):525–43 (1983).

187–188   The application of the theory of sexual selection to human sex differences is discussed in several eminently readable works: David Buss's *The Evolution of Desire* (New York: Basic Books, 1994); Donald Symons' *The Evolution of Human Sexuality* (New York: Oxford Univ. Press, 1979); and Sarah B. Hrdy's *The Woman That Never Evolved* (Cambridge, Mass.: Harvard Univ. Press, 1981). *Sex, Evolution and Behavior* by Martin Daly and Margo Wilson (Boston: Willard Grant Press, 1983) offers an authoritative, yet clear and entertaining overview of animal and human sexuality. The same authors have a short, up-to-date chapter titled "The Man who Mistook His Wife for a Chattel," pp. 289–322 in J. Barkow, L. Cosmides, and J. Tooby, editors, *The Adapted Mind* (New York: Oxford Univ. Press, 1992). For a series of detailed review articles, see L. Betzig, M. B. Mulder, and P. Turke, editors, *Human Reproductive Behavior: A Darwinian Perspective* (Cambridge: Cambridge Univ. Press, 1988).

188   For an authoritative report on male despotism and harems, see Laura L. Betzig's *Despotism and Differential Reproduction: A Darwinian View of History* (New York: Aldine, 1986).

189   The quotation from David Buss is from p. 249 in a chapter in *The Adapted Mind* (see above) on "Mate Preference Mechanisms."

189   David Buss's data are in *Behavioral and Brain Sciences*, 12:1–49 (1989). See also Bruce J. Ellis's "The Evolution of Sexual Attraction: Evaluative Mechanisms in Women" in *The Adapted Mind*, cited above.

190   The bond-testing idea is from Amotz Zahavi's "The Testing of a Bond," *Animal Behaviour*, 25:246–7 (1976).

192   Information on orgasm in primates is in Donald Symons' *The Evolution of Human Sexuality* (New York: Oxford Univ. Press, 1979).

192   For information on concealed ovulation in humans, see Beverly Strassmann's article in *Ethology and Sociobiology*, 2:31–40 (1981); Paul W. Turke's in *Ethology and Sociobiology*, 5:33–44 (1984); and Nancy Burley's in *The American Naturalist*, 114:835–58 (1979).

193   The data on testis size are from R. V. Short's chapter in *Reproductive Biology of the Great Apes*, edited by C. E. Graham (New York: Academic, 1984). See also A. H. Harcourt and collaborators' article in *Nature*, 293:55–7 (1981).

193   See R. R. Baker and M. A. Bellis's "Human Sperm Competition: Ejaculate Adjustment by Males and the Function of Masturbation," *Animal Behavior*, 46:861–85 (1993), and R. R. Baker and M. A. Bellis, "Human Sperm Competition: Ejaculation Manipulation by Females and a Function for the Female Orgasm," *Animal Behavior*, 46:887–909 (1993). Baker and Bellis's work on sperm counts is in "Number of Sperm in Human Ejaculates Varies as Predicted by Sperm Competition Theory," *Animal Behavior*, 37:867–9 (1989). For a review of work on sperm competition, see M. Gomendio and E. R. S. Roldan's "Mechanisms of Sperm Competition: Linking Physiology and Behavioral Ecology," *Trends in Ecology and Evolution*, 8(3):95–100 (1993).

194   For the work on jealousy, see Martin Daly and collaborators' "Male Sexual Jealousy," *Ethology and Sociobiology*, 3:11–27 (1982), and Martin Daly and Margo Wilson's *Homicide* (New York: Aldine, 1989). This book contains abundant data on and detailed discussion of murders motivated by jealousy.

196   For discussions of sex differences in human reproductive strategies, see the works by Buss, Ridley, Cronin, and Symons mentioned above.

197–200 David Haig's work is in *Quarterly Review of Biology*, 68:495–532 (1993). Sexually antagonistic genes are discussed by W. R. Rice in *Science*, 256:1436–9 (1992). The classic paper on parent-offspring conflict is R. L. Trivers's in *American Zoologist*, 14:249–64 (1974). A good description is also found in his book *Social Evolution* (Menlo Park, Calif.: Benjamin/Cummings, 1985). For a recent review and further references, see D. W. Mock and L. S. Forbes' article in *Trends in Ecology and Evolution*, 7(12):409–13 (1992).

200–201 The work on human birth is in a paper presented by Wenda Trevathan at the February 1993 American Academy of Sciences meeting in Boston. Also see her book *Human Birth: An Evolutionary Perspective* (Hawthorne, N.Y.: Aldine de Gruyter, 1987).

201   The work on the role of oxytocin in bonding in sheep is by E. B. Keverne et al. in *Science*, 219:81–83 (1983).

202 We got our information on the Mozarts' family tragedies mainly from pages 98–102 of *Mozart in Vienna 1781–1791* by Volkmar Braunbehrens (New York: Grove Weidenfeld, 1989).

202 On jaundice in the newborn, see John Brett and Susan Niermeyer's article in *Medical Anthropology Quarterly*, 4:149–61 (1990).

203 Defective color discrimination and other visual impairments from exposure to round-the-clock bright light in infancy are discussed by I. Abramov et al. in *Journal of the American Optometry Association*, 56:614–19 (1985).

203–204 On babies' crying, see R. G. Barr's "The Early Crying Paradox: A Modest Proposal," *Human Nature*, 1(4):355–89 (1990).

204–205 On SIDS, see James J. McKenna's "An Anthropological Perspective on the Sudden Infant Death Syndrome (SIDS): The Role of Parental Breathing Cues and Speech Breathing Adaptations," *Medical Anthropology*, 10:9–54 (1986).

205–206 On parent-offspring conflict, see the Trivers citations for pp. 195–99. Also see pp. 55–58 and 234–35 of Martin Daly and Margo Wilson's *Sex, Evolution, and Behavior*, 2nd ed. (Boston: Willard Grant Press, 1983).

Chapter 14. Are Mental Disorders Diseases?

Cases are composites to protect confidentiality.

The Moral Animal by Robert Wright (New York: Pantheon Books, 1994) offers an excellent introduction to evolutionary psychology.

A fine overview of work on evolution and psychiatry is Brant Wenegrat's *Sociobiological Psychiatry: A New Conceptual Framework* (Lexington, Mass.: Lexington Books, 1990). Forthcoming is *Evolutionary Psychiatry*, by Michael McGuire and Alfonso Troisi. For an excellent introduction to animal behavior, see John Alcock's *Animal Behavior: An Evolutionary Approach* (Sunderland, Mass.: Sinauer, 1993). For excellent introductions to sociobiology, see R. D. Alexander's *Darwinism and Human Affairs* (Seattle: University of Washington Press, 1979); R. Dawkins' *The Selfish Gene* (New York: Oxford Univ. Press, 1976); E. O. Wilson's *Sociobiology* (Cambridge, Mass.: Harvard Univ. Press, 1975); E. O. Wilson's *On Human Nature* (Cambridge, Mass.: Harvard Univ. Press, 1978); and R. Trivers' *Social Evolution* (Menlo Park, Calif.: Benjamin/Cummings, 1985). For recent progress in evolutionary psychology, see *The Adapted Mind*, cited for p. 320.

208–209 For the review that documents and emphasizes the medical orientation in current psychiatry, see Robert Michaels and Peter M. Marzuk in *New England Journal of Medicine*, 329:552–60 and 628–38 (1993).

209–212 For reviews of evolutionary approaches to emotions, see R. M. Nesse's "Evolutionary Explanations of Emotions," *Human Nature*, 1:261–89 (1990); R. Plutchik and H. Kellerman's *Theories of Emotion*, vol. 1 (Orlando, Fla.: Academic, 1980); Paul Ekman's "An Argument for Basic Emotions," *Cognition and Emotion*, 6:169–200 (1992); Robert L. Trivers's "Sociobiology and Politics," in *Sociobiology and Human Politics*, edited by E. White (Toronto: Lexington, 1981); John Tooby and Leda Cosmides's article in *Ethology and Sociobiology*, 11:375–424 (1990); R. Thornhill and N. W. Thornhill's chapter in *Sociobiology and the Social Sciences*, edited by R. Bell (Lubbock, Tex.: Texas Tech Univ. Press, 1989); and E. O. Wilson's *Sociobiology* (Cambridge, Mass.: Harvard University Press, 1975).

212 For a recent discussion on trade-offs between avoiding predation and other values, see A. Bouskila and D. T. Blumstein's article in *American Naturalist*, 139:161–76 (1992).

212–213 Walter B. Cannon's classic is *Bodily Changes in Pain, Hunger, Fear, and Rage. Researches into the Function of Emotional Excitement* (New York: Harper and Row, 1929). Also see I. M. Marks' *Fears, Phobias, and Rituals* (New York: Oxford Univ. Press, 1987); A. Öhman and U. Dimberg in *Sociopsychology*, edited by W. M. Waid (New York: Springer, 1984); I. M. Marks and Adolf Tobena in *Neuroscience and Biobehavioral Reviews*, 14:365–84 (1990); D. H. Barlow's *Anxiety and Its Disorders* (New York: Guilford, 1988); and Susan Mineka et al. in *Journal of Abnormal Psychology*, 93:355–72 (1984).

213 The fearful guppies are described by A. L. Dugatkin in *Behavioral Ecology*, 3:124–127 (1992).

213–214 For a review of signal detection theory, see D. M. Green and J. A. Swets, *Signal Detection Theory and Psycho-physics* (New York: Wiley, 1966).

214 R. H. Frank's ideas are in his book *Passions Within Reason: The Strategic Role of the Emotions* (New York: Norton, 1988).

215–216 The increasing rate of depression is documented by the Cross-National Collaborative group in "The Changing Rate of Major Depression. Cross-National Comparisons," *Journal of the American Medical Association*, 268:3098–105 (1992).

215–221 For general information on depression, see P. C. Whybrow et al. *Mood Disorders: Toward a New Psychobiology* (New York: Plenum, 1984); Emmy Gut's *Productive and Unproductive Depression* (New York: Basic Books, 1989); Paul Gilbert's *Human Nature and Suffering* (Hove, England: Erlbaum, 1989); and R. E. Thayer's *The Biopsychology of Mood and Arousal* (New York: Oxford Univ. Press, 1989).

219 The data on writers are from N. C. Andreasen's article in *The American Journal of Psychiatry*, 144:1288–92 (1987).

219 John Price's original article is in *Lancet*, 2:243–6 (1967). Also see Russell R. Gardner, Jr., in *The Archives of General Psychiatry*, 39:1436–41 (1982), and J. S. Price and Leon Sloman's article in *Ethology and Sociobiology*, 8:85s–98s (1987).

219   The data on serotonin in vervet monkeys are in M. J. Raleigh et al. article in *Brain Research*, 559:181–90 (1991).

220   For information on seasonal affective disorder, see N. E. Rosenthal and M. C. Blehar's *Seasonal Affective Disorders and Phototherapy* (New York: Guilford, 1989); D. A. Oren and N. E. Rosenthal in *Handbook of Affective Disorders*, edited by E. S. Paykel (New York: Churchill Livingstone, 1992); and David Schlager, J. E. Schwartz, and E. J. Bromet in *British Journal of Psychiatry*, 163: 322–6 (1993). The large study suggesting an increasing rate of depression is cited for p. 214.

221–222   On the studies of infant monkeys, see H. F. Harlow's *Learning to Love* (New York: Aronson, 1974).

222–224   For sources of information on attachment, see Robert Karen's review "Becoming Attached," *The Atlantic*, February 1990, pp. 35–70; John Bowlby's summary of his work in *The American Handbook of Psychiatry*, vol. 6, edited by D. D. Hamburg and H. K. H. Brodie (1969); and M. D. Ainsworth et al. *Patterns of Attachment: A Psychological Study of the Strange Situation* (Hillsdale, N.J.: Erlbaum, 1978). For a readable review of genetic focus that may influence attachment, see *Galen's Prophecy* (New York: Basic Books, 1994).

223–224   On child abuse, see Martin Daly and Margo I. Wilson's *Homicide* (New York: Aldine, 1989); their "Abuse and Neglect of Children in Evolutionary Perspective" in *Natural Selection and Social Behavior: Recent Research and Theory*, edited by R. D. Alexander and D. W. Tinkle (New York: Chiron Press, 1981); S. B. Hrdy's "Infanticide as a Primate Productive Strategy," *American Scientist*, 65:40–9 (1977); and R. J. Gelles and J. B. Lancaster, editors, *Child Abuse and Neglect* (New York: Aldine, 1987). Mark Flinn's article is in *Ethology and Sociobiology*, 9:335–69 (1988).

224–225   On schizophrenia, see J. L. Karlsson's article in *Hereditas*, 107:59–64 (1987), and J. S. Allen and V. M. Sarich's in *Perspectives in Biology and Medicine*, 32:132–53 (1988). The idea that suspiciousness may be beneficial is in a chapter by L. F. Jarvik and S. B. Chadwick in *Psychopathology*, edited by M. Hammer, K. Salzinger, and S. Sutton (New York: Wiley, 1972). For an interesting and testable idea about schizophrenia and its possible relationship to sleep cycles, see Jay R. Feierman's article in *Medical Hypotheses*, 9:455–79 (1982).

226–228   Ray Meddis's ideas are expounded mainly in his book *The Sleep Instinct* (London: Routledge and Kegan Paul, 1977); he has a shorter presentation in *Animal Behavior*, 23:676–91 (1975). For a general review of sleep among the Mammalia, see M. Elgar, M. D. Pagel, and P. H. Harvey's article in *Animal Behavior*, 40:991–5 (1990). For general reviews of sleep and sleep research, see Alexander Borbély's *Secrets of Sleep* (New York: Basic Books, 1986), and Jacob Empson's *Sleep and Dreaming* (London: Faber and Faber, 1989). For the physiology of dreaming and the possible irrelevance of psychological functions, see J. A. Hobson's *The Dreaming Brain* (New York: Basic Books, 1988); Ian Oswald, "Human Brain Proteins, Drugs, and Dreams," *Nature*, 223:893–7 (1969); and Francis Crick and Graeme Mitchison, "The Function of Dream Sleep," *Nature*, 304:111–14 (1983).

229–230   For sensorimotor constraints on dreaming, see Donald Symons' article "The Stuff That Dreams Aren't Made Of: Why Wake-State and Dream-State Sensory Experiences Differ," *Cognition*, 47:181–217 (1993).

Chapter 15. The Evolution of Medicine

The quotation at the beginning of this chapter is the title of an article by the eminent geneticist Theodosius Dobzhansky, published in *American Biology Teacher*, 35:125–9 (1973).

234–235   Readers may recognize the watch metaphor from Richard Dawkins' fine introduction to evolution, *The Blind Watchmaker* (New York: Norton, 1986). He extended the often cited idea from William Paley's 1802 masterpiece *Natural Theology*. While Paley's book was intended to clinch the case for creationism, his many examples of exquisite design provided others, including Darwin, with superb evidence for the power of natural selection. Of particular interest is Paley's attempt to explain convoluted design, which he attributes to the Deity's wish to reveal His presence to man by "contrivances" of unnecessary complexity, and by constraining His creation within the bounds of fixed laws. Paley provides a sensible view of the utility of pain but then claims that death, sickness, and their unpredictability are all necessary parts of a divinely perfect world. It was thinking of this sort that inspired Voltaire to ridicule optimists like Dr. Pangloss in his novel *Candide*.

239   For the role of antioxidants in aging, see Richard G. Cutler's "Antioxidants and Aging," *American Journal of Clinical Nutrition*, 53:373s–379s (1991). For a brief review of current research on vitamin E, see C. H. Hennekens, J. E. Buring, and R. Peto's "Antioxidant Vitamins—Benefits Not Yet Proved," *New England Journal of Medicine*, 330:1080–1 (1994).

240–241   The quote is from pp. 445–6 of René Dubos's *Man Adapting* (New Haven, Conn.: Yale Univ. Press, 1965, revised 1980).

241   The full title of Ernst Mayr's work is *The Growth of Biological Thought: Diversity, Evolution, and Inheritance* (Cambridge, Mass.: Belknap Press of Harvard Univ. Press, 1982).

241–243   Several good books address the logic of formulating questions about function, and we recommend them to those who harbor suspicions that evolutionary arguments are fundamentally illegitimate. It is a shame that such a simple misunderstanding should inhibit development of a whole field. See John Maynard Smith's *Did Darwin Get It Right?* (New York: Chapman and Hall, 1989); E. Mayr's "Teleological and Teleonomic, A New Analysis," *Boston Studies in the Philosophy of Science*, 14:91–117 (1974); John Alcock's *Animal Behavior: An Evolutionary Approach*, 4th ed. (Sunderland, Mass.: Sinauer, 1989); Michael Ruse's *The Darwinian Paradigm* (London: Routledge, 1989), George Williams' *Natural Selection* (New York: Oxford Univ. Press, 1992); and his *Adaptation and Natural Selection: A Critique of Some Current Evolutionary Thought* (Princeton, N.J.: Princeton Univ. Press, 1966).

243 The Flexner report is *Medical Education in the United States and Canada*, The Carnegie Foundation for the Advancement of Teaching, Bulletin No. 4 (1910).

248 For an informed view of the problems of modern medicine, see Melvin Konner's *The Trouble with Medicine* (London: BBC Books, 1993).

248 The article that calls for preventive health care is James F. Fries and collaborators' "Reducing Health Care Costs by Reducing the Need for Medical Services," *The New England Journal of Medicine*, 329:321–5 (1993).

INDEX